Optimization Problems in Transportation and Logistics: A Practical Guide

Optimization Problems in Transportation and Logistics: A Practical Guide

Author

Raj Bridgelall

Basel • Beijing • Wuhan • Barcelona • Belgrade • Novi Sad • Cluj • Manchester

Author
Raj Bridgelall
College of Business,
North Dakota State University
Fargo, ND,
USA

Editorial Office
MDPI
St. Alban-Anlage 66
4052 Basel, Switzerland

For citation purposes, cite as indicated below:

Lastname, Firstname, Firstname Lastname, and Firstname Lastname. Year. *Book Title*. Series Title (optional). Basel: MDPI, Page Range.

ISBN 978-3-7258-0698-0 (Hbk)
ISBN 978-3-7258-0697-3 (PDF)
doi.org/10.3390/books978-3-7258-0697-3

© 2024 by the authors. Licensee MDPI, Basel, Switzerland. This book is Open Access and distributed under the terms and conditions of the Creative Commons Attribution-NonCommercial-NoDerivs (CC BY-NC-ND) license (https://creativecommons.org/licenses/by-nc-nd/4.0/).

Table of Contents

List of Figures and Tables vii

About the Author ix

Preface xi

1. Introduction 1
   1.1. Key Concepts 1
   1.2. Overview of Methods 2
   1.3. Problem Formulation Equations 3
   1.4. Problem Formulation Table (PFT) 4
   1.5. PFT Evaluation 5
   1.6. Software Tools 5
      1.6.1. Python Program 5
      1.6.2. MIP Library 5
      1.6.3. QGIS® 6
      1.6.4. GeoDA™ 6
   1.7. Recommended Reading 6

2. Mobility Optimization 7
   2.1. Shortest Path 7
      2.1.1. Example Problem 8
      2.1.2. Solution Exercise 10
      2.1.3. Exercise on BIP with Python 10
      2.1.4. Exercise with GIS 12
      2.1.5. Further Reading 12
   2.2. Minimum Cost Tour 12
      2.2.1. Example Problem 14
      2.2.2. Solution Exercise 15
      2.2.3. Sensitivity Assessment 16
      2.2.4. Program Display 17
      2.2.5. Further Reading 18

3. Spatial Optimization 19
   3.1. Neighborhood Coverage 19
      3.1.1. Example Problem 20
      3.1.2. Solution Exercise 21
      3.1.3. Program Display 22
   3.2. Flow Capturing 22
      3.2.1. Example Problem 23
      3.2.2. Solution Exercise 24
      3.2.3. Program Display 26
      3.2.4. Further Reading 26
   3.3. Zone Heterogeneity 26

	3.3.1.	Example Problem	27
	3.3.2.	Solution Exercise	28
	3.3.3.	Program Display	29
3.4.	Heterogeneity Optimization with GIS		30
	3.4.1.	Exercise	30
	3.4.2.	Program Display	31
3.5.	Service Coverage of Locations		32
	3.5.1.	Scenario Problem with QGIS	33
	3.5.2.	Solution Exercise	34
	3.5.3.	Program Display	36

4. Spatial Logistics ... 37

- 4.1. Spatial Distribution ... 37
  - 4.1.1. Example Problem ... 38
  - 4.1.2. Solution Exercise ... 39
  - 4.1.3. Program Display ... 40
- 4.2. Flow Maximization ... 41
  - 4.2.1. Example Problem ... 42
  - 4.2.2. Solution Exercise ... 43
  - 4.2.3. Program Display ... 44
- 4.3. Warehouse Location Optimization ... 45
  - 4.3.1. Solution Exercise ... 46
  - 4.3.2. Program Display ... 47

5. Linear Programming Relaxation ... 48

- 5.1. Linear Programming Model ... 48
- 5.2. Linear Programming Library ... 49
- 5.3. Program Display ... 50

6. Summary and Conclusions ... 51

References ... 51

List of Figures and Tables

Figure 2.1:	Network representation for the shortest shipping route problem.	8
Figure 2.2:	Subtour examples that satisfy the SESX condition for all cities.	13
Figure 2.3:	Minimum tour example problem.	15
Figure 2.4:	Solution to the minimum tour example problem.	15
Figure 2.5:	Sensitivity assessment of the minimum-distance tour example problem.	16
Figure 3.1:	City neighborhood boundary for the example problem.	20
Figure 3.2:	Solution (**A**) for identical cost and (**B**) higher waterfront cost.	21
Figure 3.3:	Flow network for the example problem.	23
Figure 3.4:	Flow network solution.	25
Figure 3.5:	Neighborhood boundary map for the zone heterogeneity example problem.	28
Figure 3.6:	Solution to the zone heterogeneity example problem.	29
Figure 3.7:	Workflow depicting the use of a GIS to solve the zone heterogeneity problem.	30
Figure 3.8:	Results obtained with Texas counties using Queens contiguity weights.	32
Figure 3.9:	Solution to the service coverage example problem.	35
Figure 4.1:	Graphical representation of the spatial logistics problem.	38
Figure 4.2:	Representation of the spatial distribution solution.	40
Figure 4.3:	Solution to the example maximum flow problem.	43
Figure 4.4:	An alternate solution to the example maximum flow problem.	44
Table 1.1:	Variables and constants associated with an optimization problem.	4
Table 2.1:	PFT for the shortest path problem.	7
Table 2.2:	PFT for the shortest path problem exercise.	9
Table 2.3:	PFT for the minimum cost tour problem exercise.	14
Table 3.1:	General-form PFT for the area coverage problem.	19
Table 3.2:	PFT for the area coverage problem exercise.	21
Table 3.3:	All possible routes from start to terminal node.	24
Table 3.4:	PFT for the flow-capturing problem.	24
Table 3.5:	PFT for the zone heterogeneity problem exercise.	28
Table 3.6:	PFT for the service coverage problem.	34
Table 4.1:	Shipping cost for the spatial distribution exercise.	39
Table 4.2:	PFT for the spatial distribution exercise.	39
Table 4.3:	PFT for the maximum flow problem exercise.	42
Table 4.4:	PFT for the warehouse location problem.	46

About the Author

Raj Bridgelall holds a Ph.D. in Transportation and Logistics from North Dakota State University (USA) and master's and bachelor's degrees in Electrical Engineer-ing from Stony Brook University (USA). He has more than 140 U.S. patents issued or pending and is a fellow of the National Academy of Inventors. He has authored more than 100 peer-reviewed international publications. Professor Bridgelall teaches courses focused on technological advancements and spatial analysis in transportation and supply chain management.

Preface

Efficient transportation and logistics systems are central to global commerce and daily commuting. Hence, understanding the science behind optimizing such systems becomes indispensable. This book helps to demystify complex optimization processes while empowering students with practical skills and theoretical knowledge that are directly applicable in real-world scenarios.

I structured this book to be short and yet span a broad array of optimization problems, with a focus on linear and integer programming. The provided exercises illustrate theoretical concepts, accompanied by Python code, to help accelerate hands-on learning. Each chapter builds upon the last, presenting increasingly complex scenarios that encourage students to apply what they have learned in practical settings. An added value of this book is its incorporation of geographic information systems (GISs) in optimization problems. This modern approach allows readers to visualize the impacts of their optimization strategies geographically, which is crucial for spatial analysis in supply chain management and logistics. Students can locally execute the practical Python code examples provided by copying them from the online version of this book to reinforce learning through practical application.

For students, this book is a perfect fit for a semester-long course. I meticulously calibrated its length to match a typical academic semester. This ensures that learners can digest and engage with the material without feeling overwhelmed, attaining a sense of completion and accomplishment by the end of the course. For educators, this book provides a robust curriculum framework that they can integrate into courses related to transportation, logistics, supply chain, or any field where optimization is pertinent. The step-by-step explanations and practical examples make it an excellent reference that students can rely on long after they have completed the course.

This resource can serve as more than just a textbook—it is a toolkit for the next generation of engineers and analysts poised to innovate in the realms of transportation, logistics, and supply chain practices. It exemplifies the power of integrating theory with practical application, equipping learners with the skills necessary to tackle the challenges of tomorrow's dynamic and complex problems.

Raj Bridgelall
Author

1. Introduction

1.1. Key Concepts

Linear programming (LP) is a method used to find the optimal outcome in a mathematical model with requirements represented by linear relationships (Hillier and Lieberman 2024). The core element of LP is the objective function, which is a linear equation incorporating decision variables. The objective value is determined by the sum of products of each decision variable and its corresponding coefficient in the objective function. LP can solve problems involving the maximization or minimization of a linear objective function subject to linear equality and inequality constraints. Decision variables represent the choices available to solve a problem, such as quantities of goods to produce or transport. The objective typically involves maximizing profit or minimizing costs. LP models formulate constraints as linear equations or inequalities involving decision variables. These constraints represent the limitations or requirements of a problem, such as resource availability or demand fulfillment. Each constraint restricts the solution space to feasible solutions that adhere to all imposed limits. A fundamental theorem of LP is that a solution exists within a feasible region defined by the intersection of all constraints. This region represents all combinations of decision variables that satisfy the constraints. The optimal solution is located at a vertex of the feasible region.

The phenomenon of unbounded solutions occurs when, under certain problem formulations, the objective function can increase (in maximization problems) or decrease (in minimization problems) without bounds, indicating that no finite optimal solution exists within the defined constraints. This emphasizes the importance of carefully formulating constraints to ensure that the solution space is properly bounded. Unbounded solutions highlight a critical aspect of problem setup in LP, requiring a comprehensive understanding of both the mathematical structure and practical implications of the optimization model.

Integer programming problems fall into categories based on the nature of the decision variables. Understanding these distinct categories is crucial for selecting the appropriate solution approach. The corresponding categorization is as follows:

- Integer Linear Programming (ILP), where decision variables must take on integer values to address problems where discrete decisions are essential;
- Binary Integer Programming (BIP), a special case of ILP where variables must take on binary values (0 or 1), often representing yes/no decisions;
- Mixed Integer Programming (MIP), which allows for a mix of integer, binary, and continuous variables, accommodating a broader range of real-world problems.

This categorization facilitates a structured approach to problem solving and highlights the versatility of integer programming in tackling a broad spectrum of optimization challenges. Each category demands specific methodologies and algorithms, ranging from the precise and exact to the heuristic and approximate, to find optimal or near-optimal solutions within feasible regions defined by the constraints of each problem. As such, this guide offers a problem formulation table (PFT) that provides a structured approach to organizing and visualizing the components of an LP problem, facilitating the identification of key elements like decision variables, constraints, variable bounds, the objective function, and the appropriate class of programming.

1.2. Overview of Methods

The field of operations research encompasses a wide array of algorithms and methods designed to solve complex optimization problems that arise in various domains, including supply chain management, logistics, and transportation. The fundamental goal of these methods is to find optimal or near-optimal solutions to problems that are often characterized by a large number of variables and constraints. The following is an overview of several specialized algorithms developed to solve optimization problems; learners may consult the classic text by Hillier and Lieberman (2024) for further elaboration (Hillier and Lieberman 2024):

Simplex Method: Introduced by George Dantzig in 1947, this method is a cornerstone algorithm for solving linear programming problems. Despite its age, it remains highly effective for many applications. However, modern methods have been developed to address its inefficiencies in solving some large-scale problems.
Interior Point Methods: Developed as an alternative to the Simplex Method, this method offers a polynomial-time solution. The method is proven to be more efficient for certain large-scale optimization problems, although it may require more sophisticated implementation efforts.
Branch and Bound: Adopted for BIP and MIP formulations, this algorithm systematically explores the branches of a decision tree, which represent subdivisions of a problem, to find the optimal solution. While powerful, exponential growth of the decision tree, as occurs in large problems, can lead to computational challenges.
Genetic Algorithms and Heuristics: For problems that are NP-hard, such as many combinatorial optimization problems, exact solutions become impractical to obtain as the problem size grows. Genetic algorithms and various heuristic methods provide alternative approaches by searching for satisfactory, if not optimal, solutions within a reasonable timeframe.

Certain algorithmic approaches such as the Simplex Method or Interior Point Methods may reveal unbounded solution spaces under specific circumstances. Recognizing when an optimization problem may lead to unbounded solutions allows practitioners to refine their models or employ strategies, such as adding constraints or revising the objective function, to ensure that they can obtain meaningful, bounded solutions. Additional complexities in formulating optimization problems stem from issues regarding scalability, uncertainty, NP-hardness, and multi-objectivity. Scalability refers to the problem wherein exponentially larger computational resources are required as the size of a problem increases, making it impractical to find exact solutions within a reasonable timeframe. Uncertainty refers to the difficulty of defining variables and parameters such as those required for demand forecasting, adding complexity to a problem. NP-hardness refers to cases where there is no known polynomial-time algorithm that can solve all instances of a problem efficiently. Multi-objectivity refers to problems that involve multiple, conflicting objectives, such as minimizing cost while maximizing customer satisfaction. Overall, these complexities and challenges highlight the importance of gaining some insights into classical optimization techniques, starting with simple problems with exact solutions before tackling larger, more complex ones that may require heuristics.

1.3. Problem Formulation Equations

The meticulous definition of all the elements of an optimization problem prior to its formulation is essential for acquiring a successful solution. This preliminary step involves clearly identifying and articulating the decision variables, the objective function, constraints, and the bounds of the decision variables if any exist. Decision variables must accurately encapsulate the choices under consideration, while the objective function quantitatively reflects the goal, be it maximization of profit or minimization of cost. Constraints are vital as they delineate the feasible region by embodying the limitations or requirements inherent to a problem. Additionally, specifying bounds for the decision variables can prevent unrealistic solutions by imposing logical or physical limits. This comprehensive definition ensures that the mathematical model genuinely represents the real-world situation, thereby facilitating the development of feasible, optimal, or near-optimal solutions. Neglecting this critical preparatory phase can result in models that are either too simple or overly complex, potentially overlooking essential aspects of a problem or including superfluous elements, respectively. Thus, a thorough and precise definition of these elements, as delineated below, is essential:

(1) Define a set of decision variables; these are the choices that someone must make to achieve the optimal solution under various constraints on those choices.
(2) Define the constraints as linear functions of the decision variables with parameters defined to represent a model of the system.
(3) Define the objective function as a single linear function of the decision variables with parameters that represent one unit of the associated decision variable. Examples of parameters are cost in dollar units and flow in units of vehicles per hour.
(4) Define the lower and upper bounds of each decision variable. Binary decision variables can take on finite values of zero or one, and positive variables have a lower bound of zero.

The objective value, Z, is the dot product of the objective coefficients and the decision variables, where

$$Z = \sum_{i=1}^{N} c_i X_i \tag{1}$$

is the objective function or the functional relationship. The optimization problem consists of determining the values of X_i that will either maximize or minimize Z, subject to the constraints. The constraints can be bounds. For example, in a production optimization problem, a bound on each decision variable can be the number of items i produced, where

$$L_i \leq X_i \leq U_i \tag{2}$$

L_i and U_i are the lower and upper bounds, respectively, for item i. Additional constraints can establish the total of any items or combination of items produced.

1.4. Problem Formulation Table (PFT)

This guide introduces a cognitive framework designed to help students and practitioners formulate a problem by using a standard structure dubbed the problem formulation table (PFT). Table 1.1 shows a general structure for the proposed cognitive framework designed to help organize the facts of a problem, such as the decision variables, constraint parameters, objective function parameters, and bounds of the decision variables. This PFT is scalable so that a program can read the table as several matrices and convert the problem statement into a general form required by the selected optimizer program.

Table 1.1. Variables and constants associated with an optimization problem.

Variables (Outputs)	Constraints				Objective (Min/Max)	Variable Bounds	
	J_1	J_2	...	J_K		L	U
X_1	α_{11}	α_{12}		α_{1K}	c_1	L_1	U_1
X_2	α_{22}	α_{22}		α_{2K}	c_2	L_2	U_2
⋮	⋮	⋮	⋮	⋮	⋮	⋮	⋮
X_N	α_{N1}	α_{N2}		α_{NK}	c_N	L_N	U_N
≤	b_1	b_2		b_K	Z		

Source: Table by author.

The decision variables, X_i, are the individual outputs that the optimization program needs to find. In a production problem, the N decision variables might be individual products that a factory must produce. The decision variables have the following indices: $i = 1, 2, \ldots, N$. The K constraints can be the various resources, parts, or ingredients needed to build each product, and they have the following indices: $j = 1, 2, \ldots, K$. The fixed *parameter* values α_{ij} can represent the proportion of resource J_j needed to produce a single item, X_i. For example, α_{ij} could be the number of hours or fraction of hours needed to produce a widget or deliver a service X_i. The fixed values c_i are proportions of the objective unit associated with each decision variable X_i. For example, c_i could be the cost of producing a single widget X_i or the profit earned from selling it. The *objective coefficients* become the coefficients of the *objective function*. The last two columns of the PFT contain the lower (L) and upper (U) bounds of each decision variable. The PFT need not specify these columns when all decision variables are binary because they have the same bounds.

The constraints on resource j constitute the dot product of the resource parameter vector J_j and the decision variable vector X with an upper bound given by the value b_j in the last row. The inequality sign in the last row of the first column provides a general indication of the sign used in the constraints, separated by commas when there are several blocks of different constraints. That is, the standard formulation, as required by the optimization program, would be

$$\sum_{i=1}^{N} \alpha_{ij} X_i \leq b_j \qquad (3)$$

where b_j is the maximum amount or available capacity of resource j. The constraints for one decision variable can be a function of other decision variables. For example, if for a constraint j the quantity of item two should be twice the quantity of item one, then constraint j will be

$$X_2 = 2X_1 \qquad (4)$$

which, when converted to the *standard form*, becomes

$$-2X_1 + X_2 = 0. \qquad (5)$$

Hence, the parameter values would be $\alpha_{1j} = -2$, $\alpha_{2j} = 1$, $\alpha_{ij} = 0$, and $b_j = 0$.

1.5. PFT Evaluation

The model builder should evaluate the PFT to identify trivial infeasibilities, trivial unboundedness, and any issues in the problem formulation by examining for any of the following conditions:

1. Any column containing all zeros represents a trivial constraint.
2. A column singleton (a single 1) in an equality constraint matrix represents a fixed value for a decision variable. Hence, the model should not include this value in optimization because it is a constant, not a variable.
3. Any row containing all zeros represents an unconstrained variable.
4. A row singleton (a single 1) in an inequality constraint matrix represents a simple bound, and the model builder can use it to simplify the problem formulation.

1.6. Software Tools

1.6.1. Python Program

Python is popular in data science because of its access to a diversity of standard and specialized libraries. Python is an interpreted rather than compiled language. This means that the corresponding coding results are instantaneous because there is no need to compile code before executing it. Python is suitable for any scale project, and it is portable across many operating systems. Downloading and running the Anaconda distribution is one of the easiest ways to install Python (Anaconda Inc. 2022). This educational guide recommends using the Spyder IDE to execute the Python code examples presented. Copy and paste the source code from this document directly into the Spyder IDE code window. For all the programs in this guide, change the "datapath_in =" variable to match the location of those files on your computer. Also, ensure that the "infile" variable matches the names of the data files (Excel or CSV) that you produce when copying and pasting data tables.

1.6.2. MIP Library

Chapters 2 through 4 focus on MIP problems to demonstrate exact solutions. Chapter 5 then introduces an LP library optimized to solve problems using only continuous variables. Python has a rich ecosystem of libraries that offer several powerful tools for solving LP and MIP problems. These libraries provide a range of functionalities, spanning from simple LP solutions to complex MIP problems, catering to both beginners and advanced users in the field of operations research and optimization. For instance, the PuLP library supports a range of third-party solvers, making it very versatile and adaptable to various problem complexities. While not as comprehensive as some other libraries for complex MIP tasks, the linprog function within the SciPy library is suitable for solving small to medium-scale LP problems efficiently; hence, it is adopted in this guide for solving LP problems. In this guide, the MIP library from COIN-OR (Santos and Toffolo 2020) is employed. These tools feature high-level modeling with rapid code execution. This library is compatible with Pypy, which is a compiler that can make large MIP programs run many times faster. Learners can install the MIP library by launching the Anaconda prompt and using the PIP command as suggested in the COIN-OR documentation.

1.6.3. QGIS®

QGIS is a free open-source GIS tool that runs on many distinct types of operating systems, including Windows®, iOS®, and Linux (QGIS 2022). The standard installation offers many of the same features of popular commercial software products and many more through free open-source plugins. Follow the instructions on the QGIS website to download and install the latest stable version.

1.6.4. GeoDA™

GeoDA™ is a free software product developed by a team at the University of Chicago's Center for Spatial Data Science and partially funded by a grant from the National Science Foundation (GeoDA 2022). GeoDA focuses on tools to conduct a large variety of spatial analysis and includes the basic GIS methods necessary for performing statistical operations that involve points and polygon data. As of 2022, this tool does not support line data. Therefore, it is a complement to QGIS. One of the most notable features of GeoDA is linking, which is the ability to highlight selected data across all analysis windows. Another feature is *brushing*, which is the ability to move a selection tool in any window to dynamically highlight the selected data across maps and charts. Linking and brushing are powerful tools in exploratory spatial data analysis (ESDA) that can enable analysts to observe relationships across various visualizations of data to gain insights not otherwise possible. For example, using a box plot tool to select the outliers will highlight their locations on a map, as well as their attributes in a data table. Download and install the latest *stable* version of GeoDA from their website.

1.7. Recommended Reading

1. Shaw, Shih-Lung. 2011. Geographic information systems for transportation–An introduction. *Journal of Transport Geography* 3: 377–78. https://doi.org/10.1016/j.jtrangeo.2011.02.004 (Shaw 2011).
2. Miller, Harvey J., and Shih-Lung Shaw. 2015. Geographic information systems for transportation in the 21st century. *Geography Compass* 9: 180–89. https://doi.org/10.1111/gec3.12204 (Miller and Shaw 2015).
3. Anselin, Luc, Ibnu Syabri, and Youngihn Kho. 2006. GeoDa: an introduction to spatial data analysis. *Geographical Analysis* 38: 5–22 [pdf]. https://doi.org/10.1007/978-3-642-03647-7_5 (Anselin et al. 2010).
4. Anselin, Luc. 2017. *The GeoDA Book: Exploring Spatial Data*. Chicago: GeoDa Press LLC, Chapter 1. [pdf]. (Anselin 2017).

2. Mobility Optimization

This chapter describes how to apply the PFT introduced in Chapter 1 to set up and solve two popular optimization problems in mobility optimization: the Shortest Path and Minimum Cost Tour problems. These problems have found significant applications in areas such as transportation, logistics, network design, and electronic routing. Researchers have developed several specialized algorithms to solve these problems efficiently, each with its own strengths and applicable contexts. Examples of these specialized algorithms include Dijkstra's Algorithm, the Bellman–Ford Algorithm, and the Floyd–Warshall Algorithm. Cormen et al. (2022) provide a thorough treatment of these and other algorithms in their classic textbook (Cormen et al. 2022). In the next subsections, a PFT is used to solve these problems via ILP methods.

2.1. Shortest Path

The goal of the shortest path problem is to identify a set of links that form a path connecting a starting node s with a terminal node t such that the cost in terms of travel distance or travel time is minimal. Table 2.1 shows how to use a PFT to organize the optimization problem.

Table 2.1. PFT for the shortest path problem.

Links {L}	Node j Flow Constraint				Node j Input Constraint				Travel Time (Hours)
	1	2	...	K	1	2	...	K	
X_{11}	α_{11}	α_{12}		α_{1K}	β_{11}	β_{12}		β_{1K}	c_1
X_{21}	α_{21}	α_{22}		α_{2K}	β_{21}	β_{22}		β_{2K}	c_2
⋮	⋮	⋮	⋮	⋮	⋮	⋮	⋮	⋮	⋮
X_{KK}	α_{N1}	α_{N2}		α_{NK}	β_{N1}	β_{N2}		β_{NK}	c_N
$=, \leq$	-1	0		1	1	1		1	T

Source: Table by author.

Each decision variable X_{ij} represents a single link that exists between node i and node j. The optimization selects a link X_{ij} from a set {L} of N defined links by assigning the variable a value of 1, assigning it a value of 0 otherwise. The variable n is an index in the N links of the set {L}. The label of the starting node s is set to 1, and the label of the terminal node t is set to the total number of nodes K in the network. To represent the network digitally, the analyst must define a set of flow constraints by assigning coefficient α_{ij} a value from the set {0, 1, −1}. The flow constraint of each node j assigns $\alpha_{ij} = 1$ for all links entering the node, −1 for all links exiting the node, and 0 for all other values of α_{ij} to represent situations where there are no links between these nodes of the network. The *exit* constraint for the starting node s sums to $b_1 = -1$ because exactly one unit of flow must exit that node. Similarly, the *entry* constraint for the terminal node sums to $b_K = +1$ because the same single unit of the flow must enter node t. A *flow conservation* constraint requires the sum $b_j = 0$ ($j \neq s$, $j \neq t$) to indicate that the total units of flow entering the intermediate nodes must also exit them.

To further define the *entry* constraint, the analyst must also assign parameter β_{ij} the value 1 to specify the inputs for each node in the network. This enables the entry constraint to ensure that the unit flow enters a node from one and only one of the links connected to it. Hence, the optimization problem take the following form:

Minimize
$$T = \sum_{X_{ij} \in \{L\}}^{N} c_n X_{ij} \qquad (6)$$

subject to
$$\sum_{X_{ij} \in \{L\}}^{N} \alpha_{ij} X_{ij} = \begin{cases} -1 & \text{if } i = s \\ +1 & \text{if } i = t \\ 0 & \text{otherwise} \end{cases} \qquad (7)$$

and
$$\sum_{X_{ij} \in \{L\}}^{N} \beta_{ij} X_{ij} \leq 1 \qquad (8)$$

where
$$X_{ij} \in \{0, 1\} \qquad (9)$$

For nodes not selected on the path, the sum of input flows will be zero, and for nodes selected on the path, the sum of input flows will be exactly one, hence the reason for the $\leq$ in the second set of constraints.

2.1.1. Example Problem

The author adopted this example from a problem posed by Taylor (2019). The problem consists of finding the shortest trucking route between Los Angeles and St. Louis. Figure 2.1 shows a network representation of typical truck routes between Los Angeles and St. Louis.

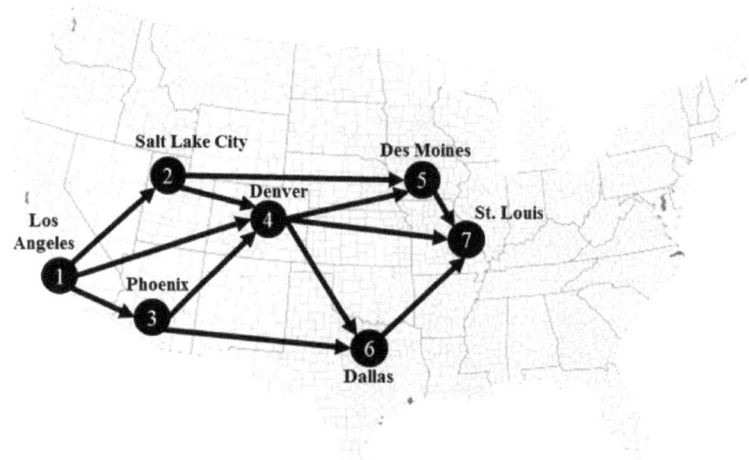

Figure 2.1. Network representation for the shortest shipping route problem. Source: Figure by author.

Table 2.2 shows the PFT designed to solve the shortest path problem using optimization software.

Table 2.2. PFT for the shortest path problem exercise.

Net Links	Node j Flow Constraints							Node j Input Constraints							Road Miles	Geodesic Miles
	1	2	3	4	5	6	7	1	2	3	4	5	6	7		
X_{12}	−1	1							1						688	579.4
X_{13}	−1		1							1					373	357.6
X_{14}	−1			1							1				1016	831.3
X_{24}		−1		1							1				519	371.8
X_{25}		−1			1							1			1066	953.1
X_{34}			−1	1							1				821	585.7
X_{36}			−1			1							1		1065	885.8
X_{45}				−1	1							1			671	610.8
X_{46}				−1		1							1		794	661.7
X_{47}				−1			1							1	851	796.7
X_{57}					−1		1							1	349	273.3
X_{67}						−1	1							1	631	547.8
=, ≤	−1	0	0	0	0	0	1	1	1	1	1	1	1	1	R	G

Source: Table by author.

The nodes represent the location of major cities, and the arcs represent established travel routes between the nodes. The first column of the PFT contains the decision variables, which are also the labels for each arc. The last two columns contain a value associated with each decision variable, which, in this case, is a distance in miles for computing the minimum total distance. The road miles provided are based on the recommended route according to Google Maps®. The Geodesic miles indicated are the geodesic distance derived using GIS software. For illustration, the trip ends are truck stop locations at the approximate center of each city. The objective function is the *dot product* of the decision variables and one of the cost columns. Starting with the road route costs, the optimization problem consists of the following:

Minimize

$$R = 688X_{12} + 373X_{13} + 1016X_{14} + 519X_{24} + 1066X_{25} + 821X_{34} \\ + 1065X_{36} + 671X_{45} + 794X_{46} + 851X_{47} + 349X_{57} + 631X_{67} \quad (10)$$

The seven node flow constraint columns define the network structure by defining the entry and exit arc flow for each node. A −1 is exit flow, and a +1 is entry flow. The last row indicates the constraint constant for the *dot product* of its respective column. In this example, the dot product must be equal to the constant. This results in the following constraints

Subject to
$$-X_{12} - X_{13} - X_{14} = -1$$

or (11)

$$X_{12} + X_{13} + X_{14} = 1$$

$$X_{12} - X_{24} - X_{25} = 0 \quad (12)$$

$$X_{13} - X_{34} - X_{36} = 0 \quad (13)$$

$$X_{14} + X_{24} + X_{34} - X_{45} - X_{46} - X_{47} = 0 \quad (14)$$

$$X_{36} + X_{46} - X_{57} = 0 \quad (15)$$

$$X_{36} + X_{46} - X_{67} = 0 \quad (16)$$

$$X_{47} + X_{57} + X_{67} = 1 \quad (17)$$

where

$$X_{ij} \in \{0, 1\} \quad (18)$$

2.1.2. Solution Exercise

The optimizer from the MIP library solves linear programming problems in the following form:

$$\min_{x} c^T x \quad (19)$$

such that

$$A^T x = b \quad (20)$$

where

$$x \in \{0, 1\} \quad (21)$$

The vector x contains the binary decision variables, c is a vector of the objective function coefficients, A is the constraint matrix extracted from the PFT, and b is a vector of the constants on the right side of the *equality* or *inequality* constraint equations. That is, each row of A^T contains the coefficients of a linear equality or inequality constraint on the decision variables in vector x. The solution corresponds to the defined order of the decision variables.

2.1.3. Exercise on BIP with Python

Create a copy of the PFT (Table 2.2) in an Excel file. Modify and run the Python program below. Change the variable "datapath_in" to match the path where the Excel file is located on your computer. Change the "infile" variable to match the name of the Excel file on your computer. The program extracts the PFT by skipping the first row to create a data frame. Because Python imports blank cells as "nan", the program replaces them with zeros. The program stores the variable names in the first column as a series of strings. As per Equation (11), the program then converts the first constraint column to positive values so that the values are binary. The program extracts the parameters into matrices and arrays that meet the MIP requirements for adding variables, objectives, and constraints to the model object.

Note that the input constraints for node $j = 1$ are all zeros, so the program eliminates these constraints when building the model. The program then calls the optimize function, which returns a status that indicates the results. The output indicates that the optimum solution is to take routes X_{14} and X_{47}, which agrees with a visual observation that it should be the shortest path. This solution results in a total of 1867 miles along primary roads. The equivalent cost in fuel consumed is the total miles divided by the average miles per gallon achieved by the truck.

```
# Author: Dr. Raj Bridgelall (raj.bridgelall@ndsu.edu)
# Shortest Path MIP
from IPython import get_ipython
get_ipython().magic('clear')                    # Clear the console
import pandas as pd
from pathlib import Path
from mip import *                               # install library from Anaconda prompt: pip install mip
import numpy as np
#%%
datapath_in = 'C:/Users/Admin/Documents/OneDrive/Documents/UGPTI/Teaching/TL 885/Lectures/6 - Mobility Optimization/Lab/'
infile = 'Shortest Path PFT.xlsx'               # Input filename
filepath_in = Path(datapath_in + infile)        # Path name for untruncated signal
df = pd.DataFrame(pd.read_excel(filepath_in, skiprows=1))   # Read Problem Formulation Table
#%%
Nc = int(np.floor(df.shape[1]/2)-1)             # Number of resource constraints
Nd = df.shape[0]-1                              # Number of decision variables
#%%
df = df.fillna(0)                               # replace all NaN (blanks in Excel) with zeros
VarName = df.iloc[0:Nd,0]                       # String of variable names
df[1] = df[1].astype(int) * -1                  # Convert starting node column to positive constraint of +1
A1_parms = df.iloc[0:Nd,1:Nc+1].astype(int)     # Extract the constraints cols
b1_parms = list(df.values[-1][1:Nc+1].astype(int))  # Get constraint parameters in an array
A2_parms = df.iloc[0:Nd,Nc+2:-2].astype(int)    # Extract the constraints cols (except 1st)
b2_parms = list(df.values[-1][Nc+2:-2].astype(int))  # Get constraint parameters in an array
c_parms = df.values.transpose()[-2][:-1].astype(int) # Get objective parameters in an array
#%%
m = Model(solver_name=CBC)                      # Instantiate the solver model object
x = [m.add_var(name=VarName[i],var_type=BINARY) for i in range(Nd)] # define decision vars in x
m.objective = minimize(xsum(c_parms[i]*x[i] for i in range(Nd)))    # add objective function
#%% Add each constraint column
for j in range(Nc):
    m.add_constr( xsum(A1_parms.iloc[i,j]*x[i] for i in range(Nd)) == b1_parms[j],\
                 "Cons1_"+str(j+1) )

for j in range(Nc-1):
    m.add_constr( xsum(A2_parms.iloc[i,j]*x[i] for i in range(Nd)) <= b2_parms[j],\
                 "Cons2_"+str(j+2) )
#%%
Status = m.optimize()
#%% Print the results
print('Model has {} vars, {} constraints and {} nzs'.format(m.num_cols, m.num_rows, m.num_nz))
selected = [VarName[i] for i in range(Nd) if x[i].x != 0]
print('Selected Paths: {} '.format(selected))
print('Minimum Cost = {} Miles'.format(m.objective_value))
for j in range(Nc+Nc-1):
    print(m.constrs[j])
print("Number of Solutions = ", m.num_solutions)
print("Status = ", Status)
```

```
The program prints the solution as follows:
Model has 12 vars, 13 constraints and 36 nzs
Selected Paths: ['X14', 'X47']
Minimum Cost = 1867.0 Miles'
Cons1_0: +1.0 X12 +1.0 X13 +1.0 X14 = 1.0
Cons1_1: +1.0 X12 -1.0 X24 -1.0 X25 = -0.0
Cons1_2: +1.0 X13 -1.0 X34 -1.0 X36 = -0.0
Cons1_3: +1.0 X14 +1.0 X24 +1.0 X34 -1.0 X45 -1.0 X46 -1.0 X47 = -0.0
Cons1_4: +1.0 X25 +1.0 X45 -1.0 X57 = -0.0
Cons1_5: +1.0 X36 +1.0 X46 -1.0 X67 = -0.0
Cons1_6: +1.0 X47 +1.0 X57 +1.0 X67 = 1.0
Cons2_0: +1.0 X12 <= 1.0
Cons2_1: +1.0 X13 <= 1.0
Cons2_2: +1.0 X14 +1.0 X24 +1.0 X34 <= 1.0
Cons2_3: +1.0 X25 +1.0 X45 <= 1.0
Cons2_4: +1.0 X36 +1.0 X46 <= 1.0
Cons2_5: +1.0 X47 +1.0 X57 +1.0 X67 <= 1.0
Number of Solutions = 1
Status = OptimizationStatus.OPTIMAL
```

2.1.4. Exercise with GIS

Use QGIS to determine the geodesic distances for the arcs between the cities. Those distances may represent more direct travel by air instead of by road. The course associated with this educational guide provides step-by-step instructions on how to complete this exercise. The last column of Table 2.2 shows the resulting values in miles. As an exercise, change the code to use those values as the new cost function, rerun the program, and compare the solutions.

2.1.5. Further Reading

1. Taccari, Leonardo. 2016. Integer programming formulations for the elementary shortest path problem. *European Journal of Operational Research* 252: 122–30. (Taccari 2016).

2.2. Minimum Cost Tour

The famous "traveling salesman" problem is a popular example in this category of problems. The problem is to determine how to visit all nodes in a network only once while minimizing the travel cost of the tour. A node on the network represents a city that the salesman must visit. The value associated with each arc is the cost of travelling between the corresponding cities. A tactic that can be used to generalize the problem of defining the network is to set an arc value to infinity, for which a direct connection between a pair of cities (points) does not exist. The mathematical formulation for the optimization problem is as follows:

Minimize

$$D = \sum_{i=1}^{N} \sum_{j=1, \ j \neq i}^{N} c_{ij} X_{ij} \qquad (22)$$

subject to

$$\sum_{i=1, \ i \neq j}^{N} X_{ij} = 1 \qquad (j = 1, 2, \ldots, N) \qquad (23)$$

and

$$\sum_{j=1, \ j \neq i}^{N} X_{ij} = 1 \qquad (i = 1, 2, \ldots, N) \qquad (24)$$

and

$$u_i - u_j + 1 \leq (N - 1)(1 - X_{ij}) \qquad (\forall i \neq 1, \forall j \neq 1) \qquad (25)$$

where

$$X_{ij} = \begin{cases} 1 & \text{the path goes from node } i \text{ to node } j \\ 0 & \text{otherwise} \end{cases} \qquad (26)$$

$$u_1 = 1, \ 2 \leq u_i \leq N \qquad (\forall i \neq 1) \qquad (27)$$

In this case, optimization entails minimizing the total distance D traveled on the tour where c_{ij} represents the distance between cities i and j. The first constraint ensures that each city j is arrived at from exactly one other city j. The second constraint ensures that the salesperson travels from each city i to exactly one other city j. Hence, the first two constraints impose a single-entry–single-exit (SESX) condition for all cities.

The SESX constraints are not sufficient for solving the full tour problem because the mathematical solution could result in one or more isolated sub-tours that meet these constraints for all cities. For example, Figure 2.2 shows a solution with subtours that meet the SESX constraint.

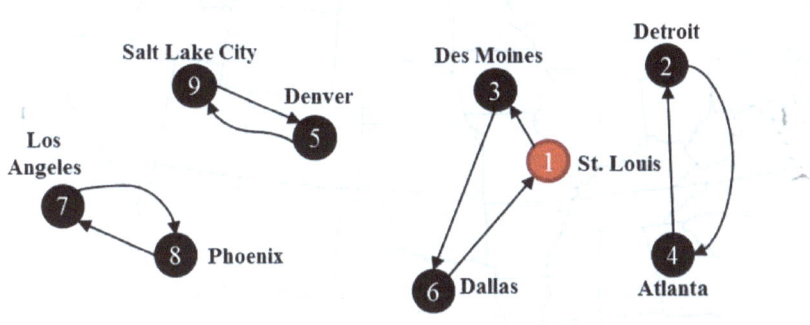

Figure 2.2. Subtour examples that satisfy the SESX condition for all cities. Source: Figure by author.

The salesperson visited all cities exactly once, but leaving any subtour to complete the full tour would violate the SESX constraint. Therefore, the formulation needs another constraint to introduce an ordering of the visits without adding new decision variables. Adding "dummy variables" u_i that are associated with each of the N network nodes results in an ordering constraint. The first dummy variable, associated with the starting node, acquires a value of 1. The other dummy variables can take on integer values between 0 and $N - 1$. The right side of the inequality becomes zero if the algorithm assigns arc $X_{ij} = 1$ because $(1 - X_{ij}) = 0$. This assignment then ensures that $(u_i - u_j) = -1$ because the left side of the inequality must become zero $(-1 + 1 = 0)$, forcing u_i to be exactly one position in the tour lower than u_j.

Table 2.3 shows the PFT for a minimum cost tour problem. There are $N \times (N - 1)$ decision variables that represent the possible number of arcs between the cities. Note that when there are more than a few nodes involved, it is not practical to use a PFT. Here, the PFT serves more as a cognitive tool with which to organize the variables, constraints, and optimization. The PFT does not include decision variables for identity nodes where $i = j$ because they are not involved in a tour. Because $u_1 = 1$ is a trivial constraint, the table includes it as a node.

Table 2.3. PFT for the minimum cost tour problem exercise.

All Links	Entry Constraints				Exit Constraints				Subtour Constraints (i, j)					Distance
	j_1	j_2	...	j_N	i_1	i_2	...	i_N	1, 2	...	1, N / N, 1	...	N − 1, N	(Miles)
X_{12}		1			1				N					c_{11}
⋮			1		1									⋮
X_{1N}				1	1									c_{1N}
X_{21}	1					1								c_{21}
⋮			1			1								⋮
X_{2N}				1		1								c_{2N}
⋮														⋮
X_{N1}	1							1						c_{N1}
⋮		1						1						⋮
$X_{N-1,N}$				1				1					N	c_{N_N-1}
u_2									1					
⋮												1		
u_N														
u_2														
⋮									−1					
u_N												−1		
=, =, ≤	1	1	1	1	1	1	1	1	N − 1				N − 1	D

Source: Table by author.

Rewriting the subtour constraint in the standard MIP format yields

$$u_i - u_j + X_{ij}N \leq (N - 1) \qquad (\forall i \neq 1, \forall j \neq 1) \qquad (28)$$

such that the right side of the inequality is a constant equal to the number of nodes visited, excluding the home node. Note that this logic holds in that if $X_{ij} = 1$, then $u_i - u_j$ must be -1 so that the inequality holds. That is, if $X_{ij} = 1$, then the equation must equal $-1 + N \leq (N - 1)$. Note that $u_i - u_j = -1$ is equivalent to $u_j = u_i + 1$. This forces the next node visited to have a dummy variable value that is higher in order by one.

2.2.1. Example Problem

The executive of a company with headquarters in Saint Louis hires a chartered flight to visit all sites where the company has a location. Figure 2.3 shows the cities that the executive must visit. The goal is to visit all the cities once and return to the headquarters in a manner that minimizes the tour cost. Hence, the chartered flight company might not visit the cities in the order of their node identifier but needs to determine the minimum distance tour.

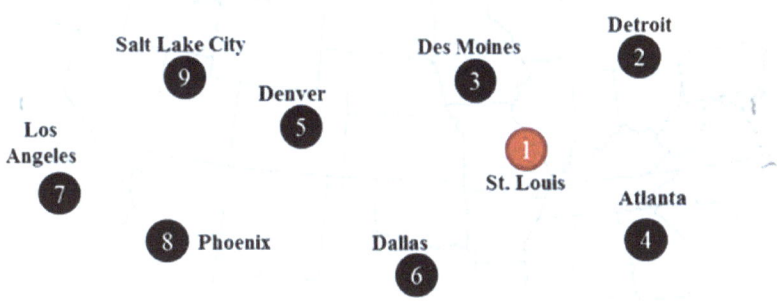

Figure 2.3. Minimum tour example problem. Source: Figure by author.

2.2.2. Solution Exercise

Use GIS to generate a distance matrix for the set of cities shown in Figure 2.3. Set the output type to linear N × 3 format and save it in a CSV file. The course associated with this educational guide provides step-by-step instructions on how to complete this exercise. Figure 2.4 illustrates the solution. The values of the u_i variables associated with each node are as shown. It is evident that the optimization assigned values that are equal to the position of the node in the tour. The minimum distance was approximately 4852 miles.

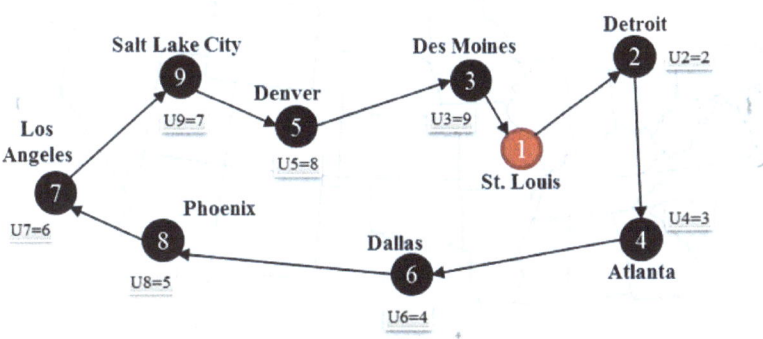

Figure 2.4. Solution to the minimum tour example problem. Source: Figure by author.

2.2.3. Sensitivity Assessment

Sometimes it is not possible to complete the tour by following the minimum-distance path. Figure 2.5 shows a scenario where the executive, while in Denver, must visit an important customer in Phoenix next. Completing the tour after this diversion results in a total distance of 5566.5 miles. Hence, as a sensitivity assessment of the minimum tour solution in this example, this diversion scenario results in a nearly 15% increase in cost.

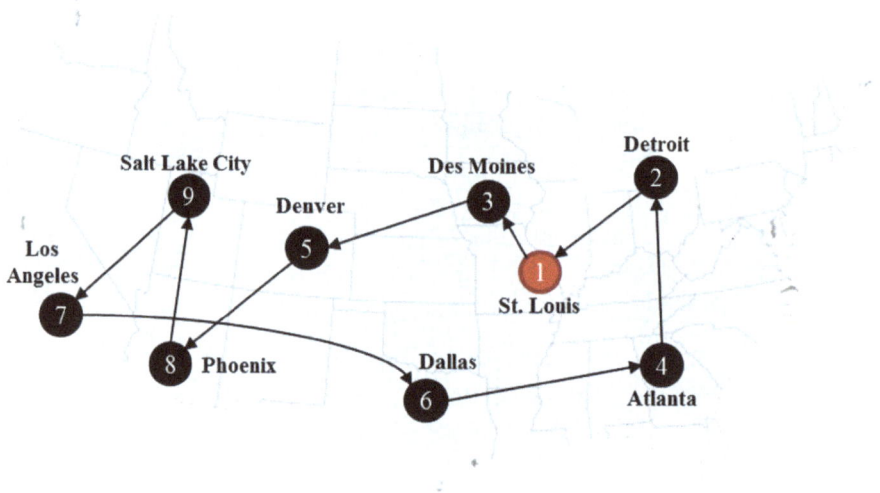

Figure 2.5. Sensitivity assessment of the minimum-distance tour example problem. Source: Figure by author.

2.2.4. Program Display

```
# Author: Dr. Raj Bridgelall (raj.bridgelall@ndsu.edu)
# Minimum Tour Optimization
from IPython import get_ipython
get_ipython().magic('clear')                           # Clear the console
get_ipython().run_line_magic('matplotlib', 'inline')   # plot in the iPython console
import pandas as pd
from pathlib import Path
from mip import *         # install library from Anaconda prompt: pip install mip
import numpy as np
import re
datapath_in = 'C:/Users/Admin/Documents/Minimum Cost Tour/Lab/'
infile = 'GDistance Matrix Nine Miles.csv'         # Input filename
filepath_in = Path(datapath_in + infile)           # Path name completion
df = pd.DataFrame(pd.read_csv(filepath_in, skiprows=0, usecols = (0, 1, 3))) # Read CSV to df
df = df.rename(columns={'InputID' : 'X[j]', 'TargetID' : 'X[i]'})    # Rename Columns
N = int(round(np.sqrt(df.shape[0]),0) + 1)         # Number of nodes to visit
N_Arcs = df.shape[0]                               # Possible arcs (distance matrix entries: N*(N-1))
#%% Create Variable Yij and add to model
VarNameX = []                                      # Initialize x[ij] variable names
VarU = []                                          # Initialize u[i] variable names
for j in range(1, N+1):
    VarU.append('U'+'_'+str(j))                    # List of dummy variable names
    for i in range(1, N+1):
        if i != j:
            VarNameX.append('X'+'_'+str(i)+'_'+str(j))  # List of decision variable names
df['Xij'] = VarNameX                               # Add variable name to table
m = Model(solver_name=CBC)                         # Instantiate optimizer
# labels and types in model vector x
x = [m.add_var(name=VarNameX[p], var_type=BINARY) for p in range(N_Arcs)]
# labels and types in model vector u
u = [m.add_var(name=VarU[q], var_type=INTEGER, lb = 2, ub = N) for q in range(1, N)]
#%% Add the cost parameters and the objective
c_parms = df.iloc[:,-2]                            # Extract distance parameters from dataframe
m.objective = minimize(xsum(c_parms[p]*x[p] for p in range(N_Arcs)))   # Objective function
#%% Single-Entry Constraint (for each j node entered, N-1 i nodes to enter from)
# Each block of j has N-1 i's AND there are N blocks of j's
for j in range(N):
    m.add_constr( xsum(x[j*(N-1) + i] for i in range(N-1)) == 1 )
#%% Single-Exit Constraint (for each i node exited, N-1 j nodes to enter)
for i in range(1, N+1):
    # Scan all variables and extract indices of all j's associated with i
    x_idx = [k for k, s in enumerate(VarNameX) if str(i) in re.split('_',s)[1] ]
    m.add_constr( xsum(x[ x_idx[j] ] for j in range(N-1)) == 1 )
#%% Sub-tour elimination constraints
for p in range(N_Arcs):
    idx_i = int(re.split('_',VarNameX[p])[1])    # Get the i index of all decision variables
    idx_j = int(re.split('_',VarNameX[p])[2])    # Get the j index of all decision variables
    if (idx_i != 1 and idx_j != 1):
        m.add_constr( u[idx_i - 2] - u[idx_j - 2] + x[p] * N <= N - 1 )   # u[0] = "U2"
Status = m.optimize()
#%% Print the results
print('Model has {} vars, {} constraints and {} nzs'.format(m.num_cols, m.num_rows, m.num_nz))
ArcName = [VarNameX[p] for p in range(N_Arcs) if x[p].x != 0]
ArcVal = [x[p].x for p in range(N_Arcs) if x[p].x != 0]
print('Tour Arcs: {} '.format(ArcName))
print('Arc Vals: {} '.format(ArcVal))
print('Minimum Total Distance = {}'.format(m.objective_value))
print("Number of Solutions = ", m.num_solutions)
print("Status = ", Status)
df['Xij_x'] = [ x[p].x for p in range(N_Arcs) ] # Add Yij and solution to the data table
print('Confirm Sum of Distance = Objective = ', sum(df[df.Xij_x != 0].Miles)) # Total distance for non-zero assignments
outfile = 'Minimum Tour.csv'                       # Table: demand sites covered by each server
filepath_out = Path(datapath_in + outfile)         # Full path name
df.to_csv(filepath_out, index = True, header = True) # Write CSV with the index column and header
```

Some portions of the code are compact and may benefit from some additional explanation. For example, the following code snippet establishes the single exit constraint:

```
#%% Single-Exit Constraint (for each i node exited, N-1 j nodes to enter)
for i in range(1, N+1):
    # Scan all variables and extract indices of all j's associated with i
    x_idx = [k for k, s in enumerate(VarNameX) if str(i) in re.split('_',s)[1] ]
    m.add_constr( xsum(x[ x_idx[j] ] for j in range(N-1)) == 1 )
```

The for loop iterates from nodes *i* ranging in index from 1 to *N*. "N + 1" is a Python standard wherein the range function returns a list of ordered values that is one less than the specified end index b in range(a, b). This makes for compact syntax, such as when the first index starts with 0, range(0, b), or simply range(b), with the last scenario returning a set of b indices starting with 0 and ending with b − 1. The optimization model stores variables in an order that starts with index 0. Hence, enumerate(VarNameX) returns an iterative array containing the strings s of the variable names and the index k in the *model* array. The logic test checks to see if the second character after the "_" string separator is the same as the variable index *i*. If so, it returns the position index k in the *model* array and builds a list of those indices as x_idx using the "list compression" feature of Python. Consequently, the final list contains the indices in the *model* of all the *j* decision variables associated with the *i* instance in question. Subsequently, constructing the exit constraint for each node *i* refers to all the *j* nodes by their decision variable x[i, j] at the position *index* stored in the model as x[index]. The regular expression split function from the re library produces an array of strings that the separator character splits. In this case, there are three strings with indices in the following range: {0, 1, 2}. The string with index 1 is the *i* value.

2.2.5. Further Reading

1. Pataki, Gábor. 2003. Teaching integer programming formulations using the traveling salesman problem. *SIAM Review* 45: 116–23. (Pataki 2003).

3. Spatial Optimization

Typical problems in planning and logistics that involve spatial optimization include neighborhood coverage, flow capturing to measure network operations, zone heterogeneity to avoid conflicts or competition, and providing full-service coverage at minimum cost. The next sections discuss each type of problem.

3.1. Neighborhood Coverage

The general problem in this category is *set covering*, where the goal is to identify a minimum set of neighborhoods to place something so that the selected locations will be adjacent to all neighboring areas. Examples include the placement of facilities such as fire stations, warehouses, and services such as emergency response centers that can serve adjacent neighborhoods within a specified amount of time. In transport logistics, sets can be neighborhoods, counties, or other areas that have boundaries. The decision variables are X_i for each area i covered, and they take on binary values. The optimization problem is expressed as follows:

Minimize

$$U = \sum_{i=1}^{N} c_i X_i \tag{29}$$

which is subject to

$$\sum_{i \in A_j} X_i \geq 1 \quad (j = 1, 2, \ldots, N) \tag{30}$$

where

$$x_i \in \{0, 1\} \tag{31}$$

Decision makers can normalize the cost to place a unit in an area by assigning a relative cost to the coefficient c_i. Table 3.1 is the PFT designed to solve the set-covering problem.

Table 3.1. General-form PFT for the area coverage problem.

Area	Area j Covers (Constraint)				Relative Cost
	1	2	...	N	
X_1	α_{11}	α_{12}		α_{1K}	1
X_2	α_{21}	α_{22}		α_{2K}	1
⋮	⋮	⋮	⋮	⋮	⋮
X_N	α_{N1}	α_{N2}		α_{NK}	1
$\geq$	1	1		1	U

Source: Table by author.

The number of areas in the "Area j Covers" constraint columns is equal to the number of areas or decision variables in the rows. Each column A_j of the constraint matrix A defines the set of areas i that area j shares a boundary with. That is, for each constraint column, the parameter is equal to 1 if area j covers area i and zero otherwise. That is, constraint column j represents the set of areas covered by a placement in area j. For example, if area 1 covers itself plus areas 2 and 3, then the first three rows of column 1 equal 1, and the remaining rows of the column equal 0. The constraint for each column is an inequality greater than or equal to 1. Hence, the dot product of the decision variables with column j must be *at least* one. That is, this constraint requires at least one placement to cover an area. The cost coefficients may be absolute cost units or relative to the minimum cost of a given neighborhood so that the objective function can make placements that minimize the total cost U to cover all areas with at least one placement.

3.1.1. Example Problem

A city with neighborhood boundaries shown in Figure 3.1 needs to build emergency response centers that can cover all neighborhoods. Decision makers determined that any station built in one neighborhood can serve neighborhoods that share its boundaries. In this scenario, the constraint must consider that the cost of building a station at the waterfront areas of 1, 4, and 7 is twice that of building elsewhere.

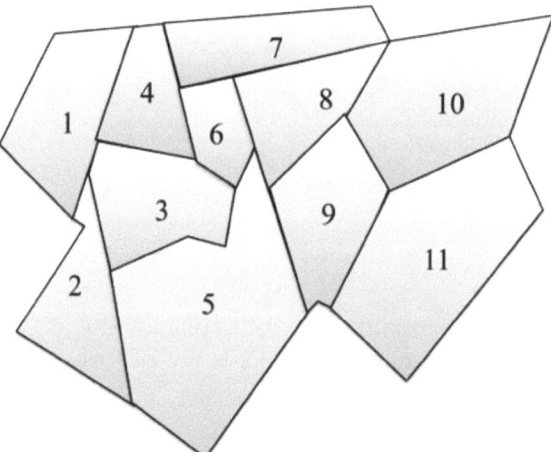

Figure 3.1. City neighborhood boundary for the example problem. Source: Figure by author.

Table 3.2 shows the PFT for the problem.

Matrix A defines the structure of the map by indicating the bordering areas for each area with a value equal to 1. Empty values are zeros.

Table 3.2. PFT for the area coverage problem exercise.

Area	Area j Covers											Cover Items (Relative Cost)
	1	2	3	4	5	6	7	8	9	10	11	
X_1	1	1	1	1								2
X_2	1	1	1		1							1
X_3	1	1	1	1	1	1						1
X_4	1		1	1		1	1					2
X_5		1	1		1	1		1	1			1
X_6			1	1	1	1	1	1				1
X_7				1		1	1	1				2
X_8					1	1	1	1	1	1		1
X_9					1			1	1	1	1	1
X_{10}								1	1	1	1	1
X_{11}									1	1	1	1
$\geq$	1	1	1	1	1	1	1	1	1	1	1	U

Source: Table by author.

3.1.2. Solution Exercise

Figure 3.2A shows the solution for a problem where the cost is the same in any area. Figure 3.2B shows the solution for when the relative cost for a waterfront area is twice the cost of any other area. The minimum number of locations is three in either case.

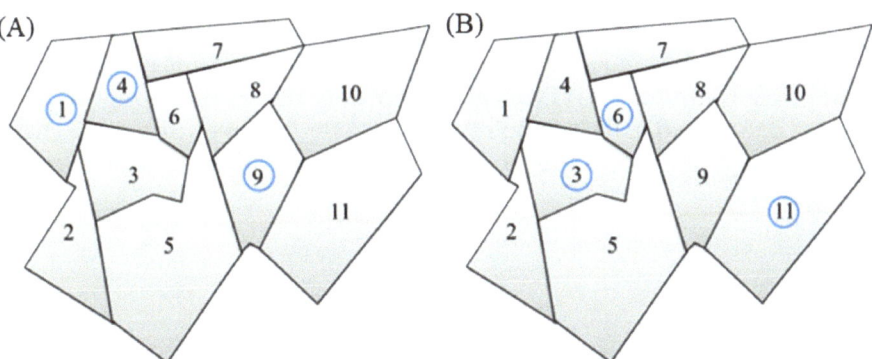

Figure 3.2. Solution (**A**) for identical cost and (**B**) higher waterfront cost. Source: Figure by author.

3.1.3. Program Display

```
# Author: Dr. Raj Bridgelall (raj.bridgelall@ndsu.edu)
# Set covering
from IPython import get_ipython
get_ipython().magic('clear')                            # Clear the console
get_ipython().run_line_magic('matplotlib', 'inline')    # plot in the iPython console
import pandas as pd
from pathlib import Path
from mip import *           # install library from Anaconda prompt: pip install mip
import numpy as np
#%%
datapath_in = 'C:/Users/Admin/Documents/Spatial Coverage Optimization/Lab/'
infile = 'Area Coverage PFT.xlsx'                   # Input filename
filepath_in = Path(datapath_in + infile)            # Path name for untruncated signal
df = pd.DataFrame(pd.read_excel(filepath_in, skiprows=1)) # Read Problem Formulation Table (PFT)
#%%
Nc = df.shape[1]-2      # Number of resource constraints
Nd = df.shape[0]-1      # Number of decision variables
#%%
df = df.fillna(0)                                   # replace all NaN (blanks in Excel) with zeros
VarName = df.iloc[0:Nd,0]                           # String of variable names
A_parms = df.iloc[0:Nd,1:Nc+1].astype(int)          # Extract the constraints cols
b_parms = list(df.values[-1][1:-1].astype(int))     # Get constraint parameters in an array
c_parms = df.values.transpose()[-1][:-1].astype(int)  # Get objective parameters in an array
#c_parms = [1 for i in range(11)]                   # Equal cost scenario
#%%
m = Model(solver_name=CBC)                          # use GRB for Gurobi
# define list of decision vars plus store in x
x = [m.add_var(name=VarName[i],var_type=BINARY) for i in range(Nd)]
m.objective = minimize(xsum(c_parms[i]*x[i] for i in range(Nd))) # add objective function
#%% Add each constraint column
for j in range(Nc):
    m.add_constr( xsum(A_parms.iloc[i,j]*x[i] for i in range(Nd)) >= b_parms[j], "Cons"+str(j) )
#%%
Status = m.optimize()
#%% Print the results
print('Model has {} vars, {} constraints and {} nzs'.format(m.num_cols, m.num_rows, m.num_nz))
selected = [VarName[i] for i in range(Nd) if x[i].x != 0]
print('Selected Areas: {} '.format(selected))
print('Minimum Cost = {} Areas'.format(m.objective_value))
for j in range(Nc):
    print(m.constrs[j])
print("Number of Solutions = ", m.num_solutions)
print("Status = ", Status)
```

3.2. Flow Capturing

This optimization problem determines the optimum set of nodes for a finite number of *placements* that will capture the maximum flow through a network. Applications in transportation include the *placement* of sensors for measuring traffic volume, weigh-in-motion (WIM) stations for measuring truck weights, driving-under-the influence (DUI) checkpoints, gasoline stations, electric vehicle charging stations, signage, and more. The problem statement is expressed as follows:

Maximize
$$F = \sum_{r \in R}^{N} f_r Y_r \tag{32}$$

which is subject to
$$Y_r \leq \sum_{j \in Y_r} X_j \quad \forall r \in R \tag{33}$$

and
$$\sum_{j=1}^{N} X_j = p \tag{34}$$

where
$$Y_r = \begin{cases} 1 & \text{if at least one placement is located on path } r \\ 0 & \text{otherwise} \end{cases} \tag{35}$$

$$X_j = \begin{cases} 1 & \text{if a placement is located at node } j \\ 0 & \text{otherwise} \end{cases} \quad (36)$$

The variable r is an index in the set of all paths R through the network from a start node to a terminal node. The variable f_r is the *maximum* flow on path r of the network. A path consists of a set of arcs in a network that is a pathway from the starting node to the terminal node. Hence, some paths may have arcs in common. In the network diagram in Figure 3.3, flows on arcs shown by dashed arrows are not part of the analysis, but they explain why flows on some arcs can be greater than the sum of flows from the arcs that connect to them. The number of available placements is p.

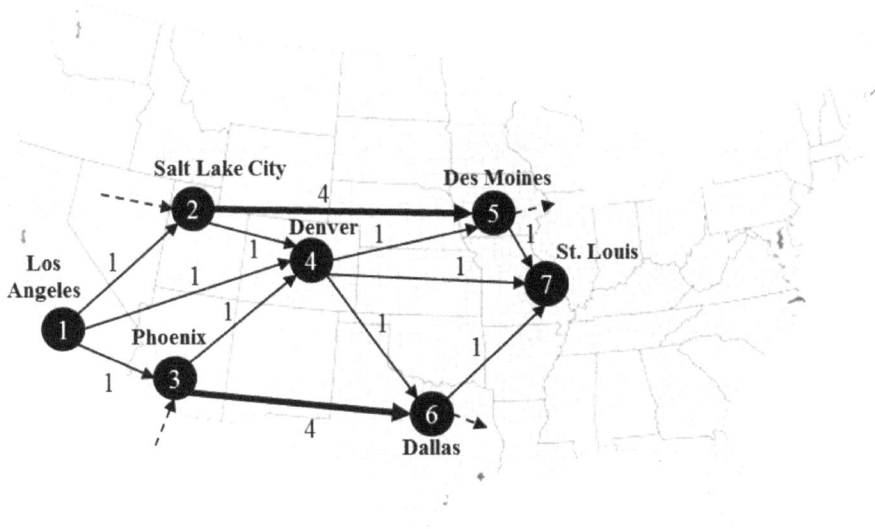

Figure 3.3. Flow network for the example problem. Source: Figure by author.

The first set of constraints along with the objective ensure that the path variable Y_r will be set to 1 if the optimization makes a placement on any node X_j along path Y_r. The index j is associated only with a node X_j that belongs to path Y_r. To enter the data into a PFT, the constraint must be in the standard form, achieved by bringing the summation to the left side of the inequality such that

$$Y_r - \sum_{j \in Y_r} X_j \leq 0 \quad \forall r \in R \quad (37)$$

The second constraint sets the available placements to the value p by summing across all N nodes of the network.

3.2.1. Example Problem

A consulting company is bidding for a grant to evaluate the performance of a new type of WIM device for the national highway system (NHS) and wants to minimize their equipment and installation budget to be cost-competitive in their proposal. Figure 3.3 shows the network on the NHS specified by the proposal solicitation. The arc labels are the *normalized* daily truck volumes. The proposal restricts placement of a WIM station at the starting and terminal nodes.

3.2.2. Solution Exercise

Table 3.3 summarizes all paths between the start and terminal nodes. Table 3.4 is the corresponding PFT. The program will delete rows containing the start and terminal nodes because these nodes cannot be candidates for a WIM station. The first constraint, labeled "p" in the header, ensures that the number of WIM stations at all nodes sums to the total p proposed. The nine path constraints Y_r encode the possible paths that exist between the start and terminal nodes, encoding the first set of constraints. The "Maximum Flow" column sets the maximum flow along each path Y_r to create the objective function. The node variables are not present in the objective function, so the flow parameters are set to zero to exclude them.

Table 3.3. All possible routes from start to terminal node.

Route	Path	Maximum Flow
Y_1	1-2-5-7	4
Y_2	1-2-4-5-7	1
Y_3	1-4-5-7	1
Y_4	1-4-7	1
Y_5	1-4-6-7	1
Y_6	1-3-4-5-7	1
Y_7	1-3-4-7	1
Y_8	1-3-4-6-7	1
Y_9	1-3-6-7	4

Source: Table by author.

Table 3.4. PFT for the flow-capturing problem.

Nodes/Paths	p	1	2	3	4	5	6	7	8	9	Maximum Flow
X_1	1	−1	−1	−1	−1	−1	−1	−1	−1	−1	0
X_2	1	−1	−1								0
X_3	1						−1	−1	−1	−1	0
X_4	1		−1	−1	−1	−1	−1	−1	−1		0
X_5	1	−1	−1	−1			−1				0
X_6	1					−1			−1	−1	0
X_7	1	−1	−1	−1	−1	−1	−1	−1	−1	−1	0
Y_1		1									4

Table 3.4. Cont.

Nodes/ Paths	p	1	2	3	4	5	6	7	8	9	Maximum Flow
Y_2			1								1
Y_3				1							1
Y_4					1						1
Y_5						1					1
Y_6							1				1
Y_7								1			1
Y_8									1		1
Y_9										1	4
$=, \leq$	3	0	0	0	0	0	0	0	0	0	F

Source: Table by author.

Run the program to acquire the solution shown in Figure 3.4 The solution consists of placing WIM stations at nodes 2, 3, and 4 to cover all the routes. The maximum flow is 15 units, as verified by the sum of flows for all paths Y_r that these nodes contain. Change the number of units to 4 and verify that the maximum flow is still 15. Change the number of units to 2 and verify that the optimization cannot achieve node placements to cover the maximum flow of 15.

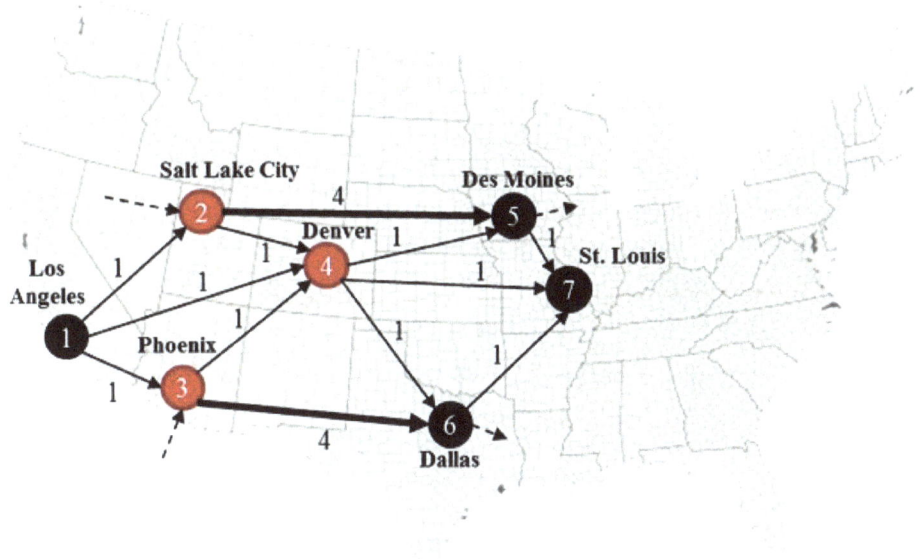

Figure 3.4. Flow network solution. Source: Figure by author.

3.2.3. Program Display

```
# Author: Dr. Raj Bridgelall (raj.bridgelall@ndsu.edu)
# Flow Capture Optimization
from IPython import get_ipython
get_ipython().magic('clear')                              # Clear the console
get_ipython().run_line_magic('matplotlib', 'inline')      # plot in the iPython console
import pandas as pd
from pathlib import Path
from mip import *          # install library from Anaconda prompt: pip install mip
import numpy as np
#%%
datapath_in = 'C:/Users/Admin/Documents/Flow Capturing/Lab/'
infile = 'Flow Capture.xlsx'                              # Input filename
filepath_in = Path(datapath_in + infile)                  # Path name for untruncated signal
df = pd.DataFrame(pd.read_excel(filepath_in, skiprows=1)) # Read Problem Formulation Table (PFT)
df = df.fillna(0)                                         # replace all NaN (blanks in Excel) with zeros
#%%
Nodes = int(df.values[df.shape[0]-1][0])                  # Get the number of inner node variables
Facilities = int(df.values[df.shape[0]-1][1])             # Get number of facilities to deploy
df = df.drop([0, Nodes+1]).reset_index(drop=True)         # Delete start and end nodes, reset table index
#%%
Nc = df.shape[1]-2      # Number of constraints
Nd = df.shape[0]-1      # Number of decision variables
#%%
VarName = df.iloc[0:Nd,0]                                 # String of variable names
A_parms = df.iloc[0:Nd,1:Nc+1].astype(int)                # Extract the constraints cols
c_parms = df.values.transpose()[-1][:-1].astype(int)      # Get objective parameters in an array
#%%
m = Model(solver_name=CBC)                                # Instantiate optimizer
x = [m.add_var(name=VarName[i],var_type=BINARY) for i in range(Nd)] # define list of decision vars plus store in x
m.objective = maximize(xsum(c_parms[i]*x[i] for i in range(Nd)))    # add objective function
#%% Add each constraint column
m.add_constr( xsum(A_parms.iloc[i,0]*x[i] for i in range(Nd)) == Facilities, "Facilities" )
for j in range(1, Nc):
    m.add_constr( xsum(A_parms.iloc[i,j]*x[i] for i in range(Nd)) <= 0, "Cons"+str(j) )
#%%
Status = m.optimize()
#%% Print the results
print('Model has {} vars, {} constraints and {} nzs'.format(m.num_cols, m.num_rows, m.num_nz))
Facility_Loc = [VarName[i] for i in range(Nodes) if x[i].x != 0]
Covered_Routes = [VarName[i] for i in range(Nodes, Nd) if x[i].x != 0]
print('Fraction of Routes Covered: {} '.format(round(len(Covered_Routes)/(Nc-1),2)))
print('Facility Locations: {} '.format(Facility_Loc))
print('Covered Routes: {} '.format(Covered_Routes))
print('Maximum Flow = {}'.format(m.objective_value))
for j in range(Nc):
    print(m.constrs[j])
print("Number of Solutions = ", m.num_solutions)
print("Status = ", Status)
```

3.2.4. Further Reading

1. Teodorović, Dušan, Milica Šelmić, Manju V. Saraswathy, Kuncheria P. Isaac, Dušan Fister, Janez Kramberger, and A. K. Sarkar. 2013. Locating flow-capturing facilities in transportation networks: a fuzzy sets theory approach. *International Journal for Traffic & Transport Engineering* 3: 103–11. (Teodorovic and Selmic 2013).

3.3. Zone Heterogeneity

Location heterogeneity means that certain aspects of neighboring locations are different. The zone heterogeneity problem involves placing items in locations whose items are unlike those of neighboring locations. The common reasons for doing so are to prevent interference, competition, an/or contamination or to induce some type of diversification. Placements can be anything from cell towers to major retail centers. A generalization of this problem is set coloring such that no adjacent area has the same color. For applications in transportation, each color could represent a facility type, service, or radio frequency channel, among other things. The last option corresponds to a frequent problem that requires determining how to select among a few available channels of narrow-band long-distance wireless communications to avoid radio-frequency interference. That is, devices in neighboring cells cannot use the same channel. In this case, unique channels are associated with unique colors of the generalized optimization problem.

The zone heterogeneity problem is similar in concept to the facility-locating problem. However, there is a significant difference in that bordering areas (neighbors) must have different or non-competing features. Hence, the solution for *neighborhood coverage*, as shown in Figure 3.2 will violate the zone heterogeneity objective because of the bordering facilities of (1, 4) and (3, 6). The problem formulation for zone heterogeneity requires defining subsets, A, that contain all area *pairs* X_{ij} that share a boundary. The optimizer sets the decision variable X_{ik} to 1 if area i is given color k. The optimization problem consist of the following: given K available colors and N nodes, minimize the number of colors assigned.

Minimize

$$Z = \sum_{i=1}^{N} \sum_{k=1}^{K} X_{ik} \tag{38}$$

which is subject to

$$\sum_{k=1}^{K} X_{ik} = 1 \quad (i = 1, 2, \ldots, N) \tag{39}$$

and

$$X_{ik} + X_{jk} \leq 1 \quad i, j \in A, \ (k = 1, 2, \ldots, K) \tag{40}$$

where

$$X_{ik} \in \{0, 1\} \tag{41}$$

The first constraint of Equation (39) establishes that the optimizer must assign every area one and only one color from K available colors. The second constraint prevents the optimizer from assigning bordering areas the same color k. The final equation is a bound specifying that the decision variables are binary.

3.3.1. Example Problem

In this example, the same neighborhood boundary map employed in the neighborhood coverage problem in Figure 3.5 is used. The decision variables represent each of the 11 areas assigned a color k. The set of areas with common borders is as follows: A = {A1_2, A1_3, A1_4, A2_3, A2_5, A3_4, A3_5, A3_6, A4_6, A4_7, A5_6, A5_8, A5_9, A6_7, A6_8, A7_8, A8_9, A8_10, A9_10, A9_11, A10_11}. Table 3.5 is the PFT encoding the facts of the problem. The 21 columns after the column of decision variables for color k represent the 21 common boundary constraints. There should be a PFT for each available color k. The program accounts for this with an iteration loop that is equal to the number of colors available. The user can change the number of available colors manually until the optimization converges by monitoring the output status of the optimizer. The PFT does not show the constraint of assigning one and only one color to an area; the program accounts for this.

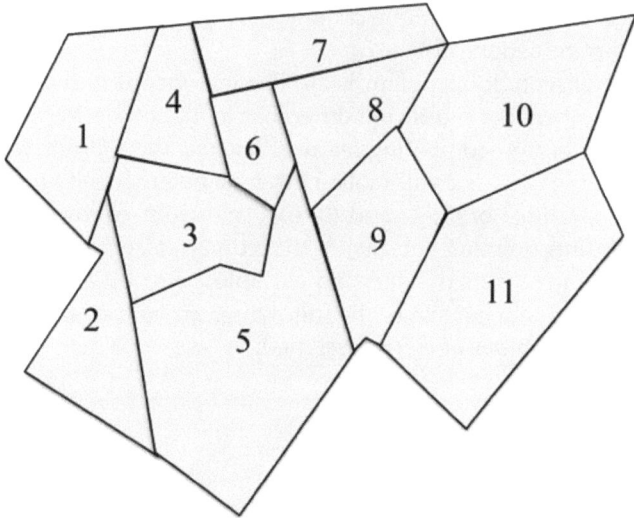

Figure 3.5. Neighborhood boundary map for the zone heterogeneity example problem. Source: Figure by author.

Table 3.5. PFT for the zone heterogeneity problem exercise.

A_{ik}	Pairwise Neighbor Constraints for One k Value																				C
	1 2	1 3	1 4	2 3	2 5	3 4	3 5	3 6	4 6	4 7	5 6	5 8	5 9	6 7	6 8	7 8	8 9	9 10	9 11	10 11	
X_{1k}	1	1	1																		1
X_{2k}	1			1	1																1
X_{3k}		1		1		1	1	1													1
X_{4k}			1			1			1	1											1
X_{5k}					1		1				1	1	1								1
X_{6k}								1	1		1			1	1						1
X_{7k}										1				1		1					1
X_{8k}												1			1	1	1				1
X_{9k}													1				1	1	1		1
X_{10k}																		1		1	1
X_{11k}																			1	1	1
$\leq$	1	1	1	1	1	1	1	1	1	1	1	1	1	1	1	1	1	1	1	1	Z

Source: Table by author.

3.3.2. Solution Exercise

Figure 3.6 shows a solution that uses four colors. The colors represent categorical values and have no ordinal meaning. Users may select a suitable color scheme from the website ColorBrewer.org (Brewer and Harrower 2020).

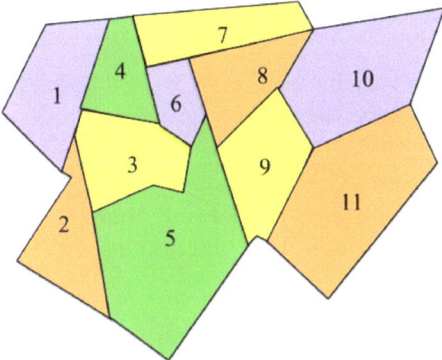

Figure 3.6. Solution to the zone heterogeneity example problem. Source: Figure by author.

3.3.3. Program Display

```
# Author: Dr. Raj Bridgelall (raj.bridgelall@ndsu.edu)
from IPython import get_ipython
get_ipython().magic('clear')                            # Clear the console
get_ipython().run_line_magic('matplotlib', 'inline')    # plot in the iPython console
import pandas as pd
from pathlib import Path
import numpy as np
from mip import *
#%%
datapath_in = 'C:/Users/Admin/Documents/Spatial Heterogeneity Optimization/Lab/'
infile = 'Color Map PFT.xlsx'                           # Input filename
filepath_in = Path(datapath_in + infile)                # Path name for untruncated signal
df = pd.DataFrame(pd.read_excel(filepath_in, skiprows=1))   # Problem Formulation
#%%
Nc = df.shape[1]-2      # Number of boundary constraints per color
Nd = df.shape[0]-2      # Number of area decision variables per color
Ncolor = 4              # Number of colors available (modify until optimization converges)
#%%
df = df.fillna(0)                           # replace all NaN (blanks in Excel) with zeros
#VarName = df.iloc[1:Nd+1,0]                # String of variable names
VarName = list(df.iloc[1:Nd+1,0])           # List of variable name strings for manipulation
A_parms = df.iloc[1:Nd+1,1:Nc+1].astype(int)    # Extract the constraints cols
#%%
m = Model(solver_name=CBC)                  # use GRB for Gurobi
#%% decision variable is whether or not to color area i with color k: create all combination
x = [m.add_var(name=VarName[i]+'_'+str(k), var_type=BINARY) for k in range(Ncolor) for i in range(Nd) ]
#%% Minimize total number of colors assigned (subject the constraints)
m.objective = minimize(xsum(x[i] for i in range(Nd*Ncolor)))  # add objective function
#%% For each color, add constraints for each boundary constraint column
for k in range(Ncolor):
    for j in range(Nc):
        m.add_constr( xsum(A_parms.iloc[i,j]*x[k*Nd+i] for i in range(Nd)) <= 1,
"Cons_jk_"+str(j)+'_'+str(k) )
#%% Add constraint for a single color per area
for i in range(Nd):
    m.add_constr( xsum(x[k*Nd+i] for k in range(Ncolor) ) == 1 )
#%%
Status = m.optimize()
#%% Print the results
print('Model has {} vars, {} constraints and {} nzs'.format(m.num_cols, m.num_rows, m.num_nz))
print('Objective Function: \n',m.objective)
print('Constraints:')
for j in range(m.num_rows):
    print(m.constrs[j])
print("Status = ", Status)
print("Number of Solutions = ", m.num_solutions)
selected = [str(x[i]) for i in range(Nd*Ncolor) if x[i].x != 0]
print('Areas Colored: {} '.format(selected))
for i in range(Nd*Ncolor):
    print('{}: {} = {}'.format(i,str(x[i]),x[i].x))
#%% Number of times a color is used
C = np.zeros(Ncolor)
for k in range(Ncolor):
    for i in range(Nd):
        if x[k*Nd+i].x != 0:
            C[k] = C[k] + 1
for k in range(Ncolor):
    print('Color {} Used: {} '.format(k,C[k]))
```

The program adds constraints for each color by using the loop iterative for k in range(Ncolor):

Indexing the decision variable as x[k*Nd + i] automatically selects the variable block for each color in the loop iteration.

3.4. Heterogeneity Optimization with GIS

This section demonstrates how to write a Python program that creates decision variables, constraints, and the objective function directly from the output of a GIS. Figure 3.7 shows the workflow.

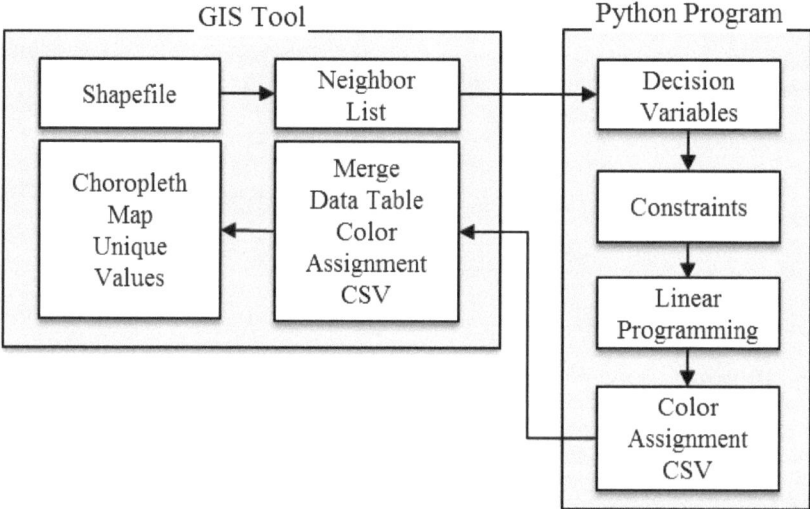

Figure 3.7. Workflow depicting the use of a GIS to solve the zone heterogeneity problem. Source: Figure by author.

3.4.1. Exercise

The course associated with this educational guide provides a step-by-step guide on how to execute this workflow. The general steps are as follows:

1. Use GeoDA to import a shapefile containing all the counties of a selected state.
2. Use the GeoDA Weights Manager tool to find all the neighbors of each county and store the output in a weights file. Use the Queens weighting criteria.
3. Run the Python program in the next section to solve the spatial optimization problem and produce a CSV file of the color category assigned to each area. Change the datapath_in and infile variables so that they match the location and filename on your machine.
4. Merge the color assignment table with the map data table using the unique code assigned for each county.
5. Generate a categorical choropleth map based on the unique color values.

GeoDA defines "Rook" and "Queens" contiguity to determine the criteria for neighboring polygons based on the movement rules of the game of chess. Neighboring polygons share only lines in the Rook criteria, whereas neighbors also share vertices in the Queens criteria. Figure 3.8 shows the results regarding Texas counties using Queens contiguity weights.

3.4.2. Program Display

```python
# Author: Dr. Raj Bridgelall (raj.bridgelall@ndsu.edu)
# Spatial Optimization (Color Maps) using GeoDA output
from IPython import get_ipython
get_ipython().magic('clear')                              # Clear the console
get_ipython().run_line_magic('matplotlib', 'inline')      # plot in the iPython console
import pandas as pd                                       # Data wrangling
from pathlib import Path                                  # File path management
import numpy as np                                        # Numerical library
from mip import *                                         # Multiple Integer Programming Library
import re                                                 # Regular Expression string processing
#%% Read in the .gal file from GeoDA
datapath_in = 'C:/Users/Admin/Documents/Zone Heterogeneity/Lab/'
infile = 'Color State_TX_Queen.gal'           # Input file from GeoDA Weights (Neighbors)
filepath_in = Path(datapath_in + infile)      # Full path name for file
list_of_lists = []                            # Initialize file input list
with open(filepath_in) as f:                  # Open the file as object f
    for line in f:                            # Get one line at a time
        inner_list = [line.strip() for line in line.split(' ')]   # Split line by space dilimeter
        list_of_lists.append(inner_list)                          # Append row of string values
#%%
Ncolor = 4    # Number of available colors--increase manually from 2 until optimization converges.
N_items = int(list_of_lists[0][1])  # First line second value = totol number of areas in the map
#%% Get the decision variable names (one per item for a given color)
VarName = []                        # Initialize the list
for i in range(1,N_items*2,2):  # Start at line 1, loop every other line ['Area', 'Num Neighbors']
    VarName.append(list_of_lists[i][0])      # Extract variable name string
#%% Instantiate the MIP solver
m = Model(solver_name=CBC)                                # use GRB for Gurobi
#%% decision variable is whether or not to color area i with color k: create all combination.
# Use '_' as i_k separator. Store in object list x (optimizer refer it it only)
x = [m.add_var(name=VarName[i]+'_'+str(k), var_type=BINARY) for k in range(Ncolor) for i in
range(N_items) ]
#%% Minimize total number of colors assigned (subject the constraints)
m.objective = minimize(xsum(x[i] for i in range(N_items*Ncolor)))    # Objective function
#%% For each color, add constraints for each neighbor pair in the .gal file
for k in range(Ncolor):
    for i in range(1,N_items*2,2):
        Var1 = x[VarName.index(list_of_lists[i][0]) + k*N_items]     # Get x[i] at VarName index
        for j in range(int(list_of_lists[i][1])):  # Alternate lines = ['Area', 'Num Neighbors']
            Var2 = x[VarName.index(list_of_lists[i+1][j]) + k*N_items]   # Get x[j] on next line
            m += Var1 + Var2 <= 1        # Add the x[i_k] + x[j_k] <= 1 constraint to the model
#%% Add the constraint that each area must have one and only one color
for i in range(N_items):
    m.add_constr( xsum(x[k*N_items+i] for k in range(Ncolor) ) == 1 ) # sum x[i] for all k = 1
#%% Run the optimizer and return the status
Status = m.optimize()
#%% Print the results
print('Model has {} vars, {} constraints and {} nzs'.format(m.num_cols, m.num_rows, m.num_nz))
print("Status = ", Status)
print("Number of Solutions = ", m.num_solutions)
selected = [str(x[i]) for i in range(N_items*Ncolor) if x[i].x != 0]  # Variables assigned a color
print('Areas Colored: {} '.format(selected))
#%%
ItemID = [re.split('_',s) for s in selected]        # Get list of area names and color assigned
ItemDF = pd.DataFrame(ItemID)                       # Convert to data frame for CSV export
ItemDF = ItemDF.rename(columns = {0: 'County_FP', 1: 'Color'})  # Name the columns in the header
outfile = 'Color State_WY.csv'                      # Output CSV filename
filepath_out = Path(datapath_in + outfile)          # Full file name including Path
ItemDF.to_csv(filepath_out, index = False, header = True)   # Write CSV without the index column
#%% Number of times a color is used
C = np.zeros(Ncolor)                                # Initialize the color counter array
for k in range(Ncolor):
    for i in range(N_items):
        if x[k*N_items+i].x != 0:       # If a color is assigned, accumulate color count for k
            C[k] = C[k] + 1
for k in range(Ncolor):                             # Print number of time each color was used
    print('Color {} Used: {} '.format(k,C[k]))
```

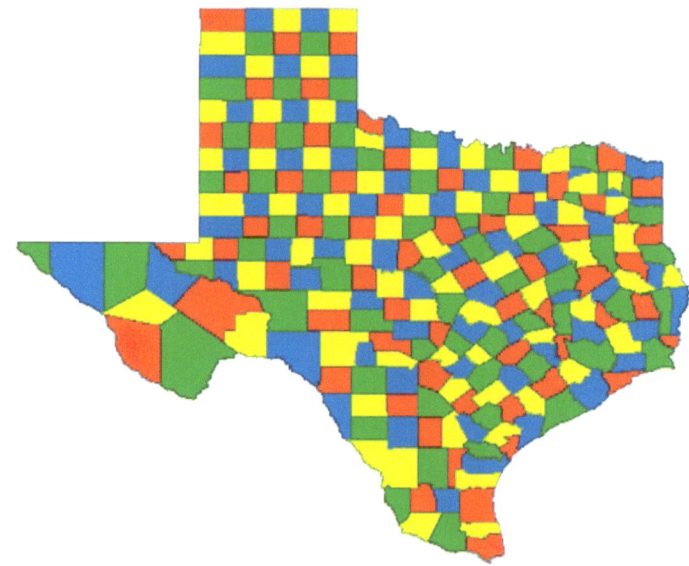

Figure 3.8. Results obtained with Texas counties using Queens contiguity weights. Source: Figure by author.

3.5. Service Coverage of Locations

This problem relates to the distribution of a finite number of *servers*, such as gateways, sensors, vehicles, facilities, and services, that can serve a finite set of demand locations. The problem concerns where to place servers among a set of candidate locations. The same server can service multiple demand locations, but no more than one server can service the same demand location. The optimization problem entails distributing the servers in a manner that minimizes the total cost to serve all demand locations. Cost can be distance, travel cost, wireless transmission cost, and other possibilities for minimization. Using distance as a direct cost, the variables are as follows:

I the set of N *demand* node locations indexed by I;
J the set of M candidate server locations indexed by j;
p the number of servers available;
d_{ij} the physical distance between demand node i and candidate server location j.

The problem is formulated as follows:

Minimize

$$D = \sum_{i=1}^{N}\sum_{j=1}^{M} d_{ij} Y_{ij} \qquad (42)$$

which is subject to

$$\sum_{j=1}^{M} Y_{ij} = 1, \forall i \in I \qquad (43)$$

and

$$\sum_{j=1}^{M} X_j = p \qquad (44)$$

and
$$Y_{ij} \leq X_j, \forall i \in I, \forall j \in J \tag{45}$$

where

$$Y_{ij} = \begin{cases} 1 & \text{location } i \text{ is served from location } j \\ 0 & \text{otherwise} \end{cases}, \forall i \in I, \forall j \in J \tag{46}$$

$$X_j = \begin{cases} 1 & \text{if server is placed at location } j \\ 0 & \text{otherwise} \end{cases}, \forall j \in J \tag{47}$$

The objective function selects candidate sites that minimize the overall service distance in the network. The first constraint ensures that one and only one server will serve a demand site. The second constraint ensures that the number of server placements is exactly p. The third constraint ensures that if the optimizer places a server at location j to serve location i, then it must set the location of server j to assigned. All the decision variables are binary.

3.5.1. Scenario Problem with QGIS

Many cities have been considering using land use planning to accommodate micromobility. During the early deployment of micromobility vehicles such as electric scooters and electric bicycles, cities faced many challenges (NACTO 2018). The convenience of stowing a vehicle anywhere resulted in clutter and interference with pedestrian traffic. Consequently, cities have been restricting the use of undocked micromobility vehicles. Meanwhile, companies have proposed new types of micromobility vehicles, including some that have an enclosed compartment to provide heat and shelter. Cities recognize that the demand for such vehicles will increase because they provide a lower-cost alternative for short trips, and they can connect to public transportation for trip completion or longer trips. Hence, affordable and accessible micromobility devices could help relieve traffic congestion, particularly if deployment results in a mode shift away from single-occupancy vehicles.

The problem scenario posed for this exercise is as follows: one city proposed to design small, marked spaces next to bus stops and subway stations where users can safely stow micromobility vehicles after use or incur a penalty if they do not. The pilot study will deploy a fleet of micromobility vehicles near a popular park in a densely populated suburb. Planners wish to start with three bus stops selected from nine candidate locations and the subway stations located within four kilometers of the park centroid. Their objective is to minimize the total distance between the selected locations and the subway stations. Table 3.6 shows the PFT for setting up the optimization problem using software.

Table 3.6. PFT for the service coverage problem.

Service Links	p			Service Constraints for j					Distance Matrix
		0	1	...	M	1	...	M	
Y_{11}	0		1			1			d_{11}
$\vdots$	0			1			1		$\vdots$
Y_{NM}	0				1			1	d_{NM}
X_1	1				−1				0
$\vdots$	1						−1		$\vdots$
X_M	1							−1	0
=, ≤	p	1	1	1	0	0	0	0	D

Source: Table by author.

3.5.2. Solution Exercise

Figure 3.9 illustrates the solution.

The course associated with this educational guide provides step-by-step instructions on how to complete this exercise. The general workflow is as follows:

(1) Download shapefiles for New York City subway and bus stops.
(2) Use QGIS to select all subway and bus stops that are within 4 km from the centroid of Prospect Park in Brooklyn, New York.
(3) Use QGIS to select nine candidate sites at bus stops around the park.
(4) Use QGIS to produce a distance matrix for all combinations of distances between the candidate sites and the subway stations.
(5) Save the distance matrix as a CSV file.
(6) Run the Python program to use the distance matrix as an input file, build the optimization model, and write the results into a cover map file of the subway stations covered by the selected bus-stop sites.
(7) Use QGIS to display a color-coded map of the results.

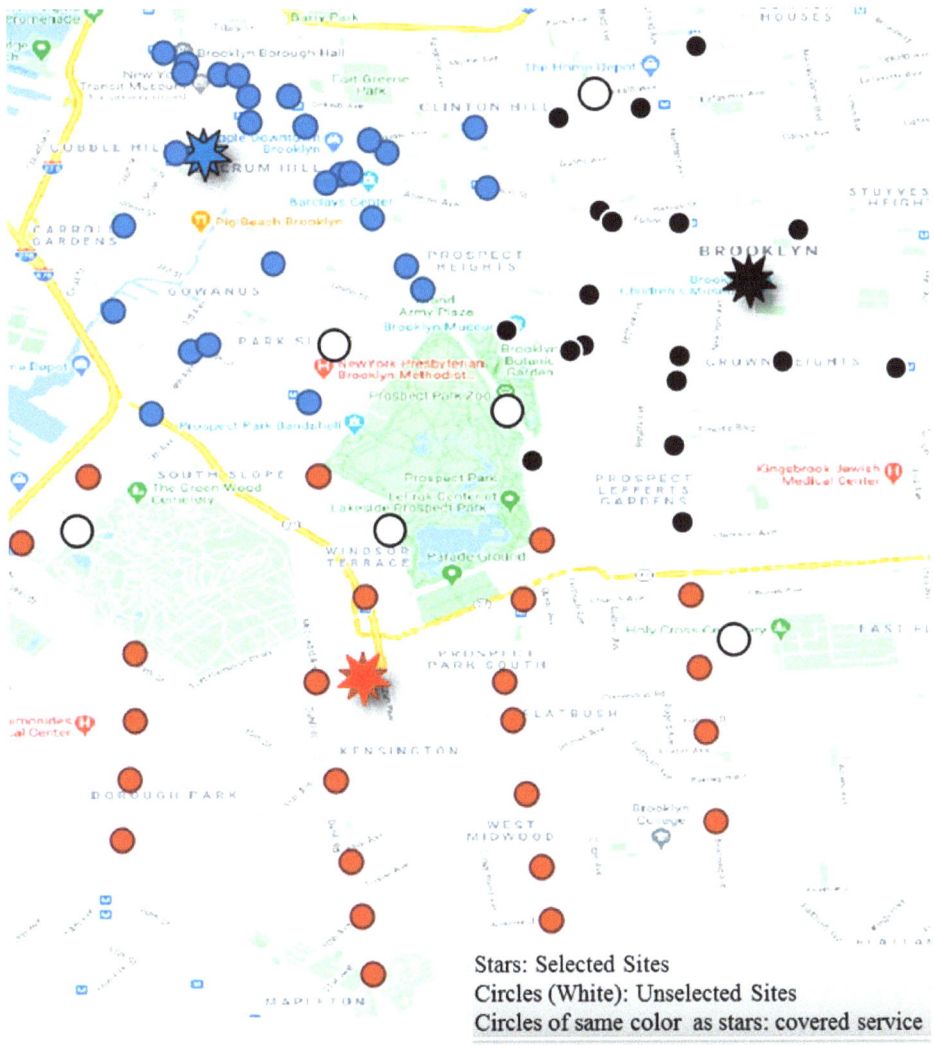

Figure 3.9. Solution to the service coverage example problem. Source: Figure by author.

3.5.3. Program Display

```python
# Author: Dr. Raj Bridgelall (raj.bridgelall@ndsu.edu)
# Service Coverage Optimization
from IPython import get_ipython
get_ipython().magic('clear')                              # Clear the console
get_ipython().run_line_magic('matplotlib', 'inline')      # plot in the iPython console
import pandas as pd
from pathlib import Path
from mip import *          # install library from Anaconda prompt: pip install mip
import numpy as np
#%%
datapath_in = 'C:/Users/Admin/Documents/Land Use Modeling/Lab/'
infile = 'Distance Matrix Bus Subway.csv'      # Input filename
filepath_in = Path(datapath_in + infile)       # Path name completion
df = pd.DataFrame(pd.read_csv(filepath_in, skiprows=0)) # Read CSV into dataframe
df = df.rename(columns={'InputID' : 'BusStop[j]', 'TargetID' : 'Subway[i]'})   # Rename Columns
#%%
N_Servers = 3                        # Number of serving stations
M_BusStop = 9                        # Number of target sites for serving stations
N_Subways = int(df.shape[0]/M_BusStop)  # Number of demand sites to serve
MN = df.shape[0]                     # Total combination of MN for loops
#%% Create Variable Yij and add to model
VarNameY = []                                             # Initialize list
for j in range(1, M_BusStop+1):
    for i in range(1, N_Subways+1):
        VarNameY.append('Y'+str(i)+'_'+str(j))            # Create list of decision variable names
m = Model(solver_name=CBC)                                # Instantiate optimizer
y = [m.add_var(name=VarNameY[p], var_type=BINARY) for p in range(MN)] # y = y vars
#%% Add the cost parameters and the objective
c_parms = df.iloc[:,-1]                                   # Extract distance parameters from dataframe
m.objective = minimize(xsum(c_parms[p]*y[p] for p in range(MN)))    # objective function
#%% constraint for demand i serviced by only one server j
for i in range(N_Subways):
    m.add_constr( xsum(y[j*N_Subways + i] for j in range(M_BusStop)) == 1 )
#%% Create Variable Xi and add to model
VarNameX = []
for i in range(1, N_Subways+1):
    VarNameX.append('X'+str(i))
x = [m.add_var(name=VarNameX[i], var_type=BINARY) for i in range(N_Subways)]   # x = x vars
#%% Add constraint Yij - Xj <= 0
for j in range(M_BusStop):
    for i in range(N_Subways):
        m.add_constr( y[j*N_Subways + i] - x[j] <= 0 )   # y has M blocks of i (each N long)
#%% Add constraint sum(xj) = p
m.add_constr( xsum( x[j] for j in range(M_BusStop) ) == N_Servers )
#%%
Status = m.optimize()
#%% Print the results
print('Model has {} vars, {} constraints and {} nzs'.format(m.num_cols, m.num_rows, m.num_nz))
Server_Loc = [VarNameX[i] for i in range(M_BusStop) if x[i].x != 0]
Service_Net = [VarNameY[p] for p in range(MN) if y[p].x != 0]
print('Server Locations: {} '.format(Server_Loc))
print('Service Network: {} '.format(Service_Net))
print('Minimum Total Distance = {}'.format(m.objective_value))
print("Number of Solutions = ", m.num_solutions)
print("Status = ", Status)
#%% Add Yij variables and solution to the data table
df['Yij'] = [ VarNameY[p] for p in range(MN) ]
df['Yij_x'] = [ y[p].x for p in range(MN) ]
print('Confirm Sum of Distance = Objective = ', sum(df[df.Yij_x != 0].Distance)) # NZ tot dist
#%% Add Xj variables and solution to the data table
# Table has M blocks of i (each N long)
df['Xj'] = [ VarNameX[j] for j in range(M_BusStop) for i in range(N_Subways) ]
df['Xj_x'] = [ x[j].x for j in range(M_BusStop) for i in range(N_Subways) ]     # matching values
#%% Table of demand sites covered by each server
CoverMap = df.pivot_table(values='Yij_x', index = ['Subway[i]'], columns=['BusStop[j]'], aggfunc=np.sum)
CoverTable = pd.DataFrame(CoverMap.sum().rename('Served'))   # See in 'Variable Explorer'
outfile = 'Cover Table.csv'                               # Output filename
filepath_out = Path(datapath_in + outfile)                # Full path name
CoverTable.to_csv(filepath_out, index = True, header = True)  # Write CSV w/ index col & header
#%% List of demand sites covered by servers
ColorMap = pd.get_dummies(CoverMap).idxmax(1)             # Reverse one-hot-encoding
ColorMap = ColorMap.rename('BusStop[j]')                  # Rename column
outfile = 'Subway Color Map.csv'                          # Output filename
filepath_out = Path(datapath_in + outfile)                # Full path name
ColorMap.to_csv(filepath_out, index = True, header = True)  # Write CSV w/ index col & header
```

4. Spatial Logistics

Problems in spatial logistics involve distributing items to satisfy supply and demand constraints while minimizing cost or maximizing flow. Hence, in the literature, this goal is also known as the transport problem (Hillier and Lieberman 2024). The next sections cover three types of spatial logistics problems that involve distribution, flow, and warehouse location optimization.

4.1. Spatial Distribution

This category of optimization problems involves decision making to meet supply and demand constraints. The decision variables X_{ij} are associated with each combination of supply entity i and demand entity j. The objective function is supplier cost minimization, where the coefficient c_{ij} represents the per-unit cost of production plus transportation. The problem formulation for a single item from M suppliers and N demand locations is as follows:

Minimize

$$Z = \sum_{i=1}^{M} \sum_{j=1}^{N} c_{ij} X_{ij} \qquad (48)$$

which is subject to

$$\sum_{i=1}^{M} X_{ij} = d_j (j = 1, 2, \ldots, N) \qquad (49)$$

and

$$\sum_{j=1}^{N} X_{ij} \leq \sum_{j=1}^{N} d_j (i = 1, 2, \ldots, M) \qquad (50)$$

where

$$X_{ij} \geq 0 (i = 1, 2, \ldots, M), (j = 1, 2, \ldots, N) \qquad (51)$$

The first constraint of Equation (49) is a *demand* constraint ensuring that the solution meets the demand for all locations. It states that the total units supplied from all suppliers i to location j must be equal to the demand from location j. The second constraint is a *supply* constraint that establishes an upper bound on supplier i. It states that the total amount from supplier i can be at most the total demand from all locations j. This constraint determines the maximum design capacity for each supplier i when the logistics cost of the *system* is minimal. An alternate supply constraint consists of setting the existing capacity of each supplier as

$$\sum_{j=1}^{N} X_{ij} \leq U_i (i = 1, 2, \ldots, M) \qquad (52)$$

The variable bound Equation (51) establishes a non-negative constraint on the quantity of items supplied. That is, the decision variables are *integers*, and they have only a lower bound.

4.1.1. Example Problem

The author adopted this example from a popular example involving warehouse suppliers and stores (Mitchell et al. 2020). In this scenario, the supplier has two warehouses that supply smartphones to five stores in the region. Each week, the average demand from stores 1 through 5 is 500, 900, 1800, 200, and 700 units, respectively. Figure 4.1 is a graphical representation of the problem.

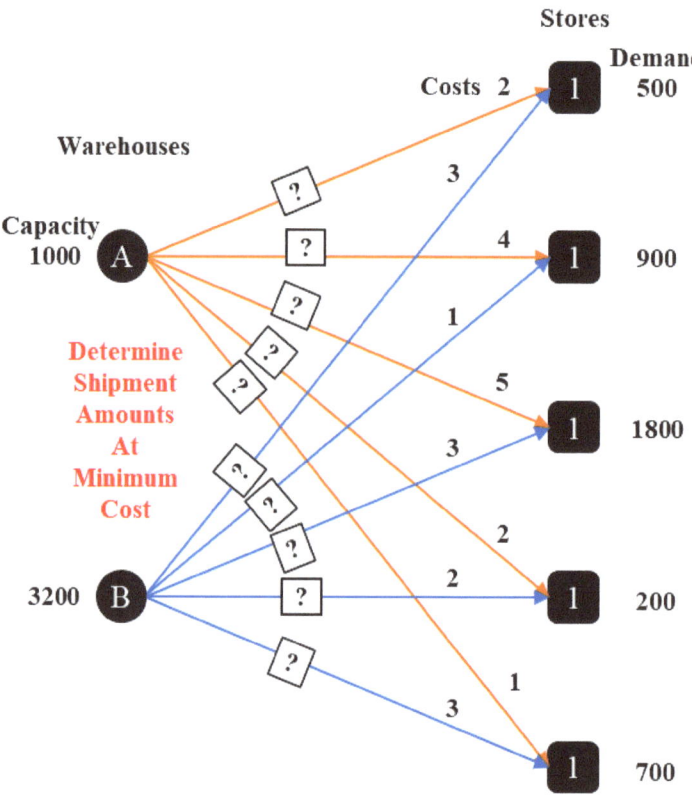

Figure 4.1. Graphical representation of the spatial logistics problem. Source: Figure by author.

One week, the warehouse manager noted that they could supply only 1000 units from warehouse A and 3200 units from warehouse B. Given that the cost per unit shipment depends on the truck route between a given warehouse and a store (Table 4.1), the manager needed to determine the optimum distribution of shipments from each warehouse to stores that will minimize the shipping costs while meeting the demand.

Table 4.1. Shipping cost for the spatial distribution exercise.

To Store	From Warehouse	
	A (USD)	B (USD)
1	2	3
2	4	1
3	5	3
4	2	2
5	1	3

Source: Table by author.

Table 4.2 shows the PFT for this spatial distribution problem. The five demand constraints are equalities (eq), and the two supply constraints are upper bounds (ub).

Table 4.2. PFT for the spatial distribution exercise.

Ship	Stores (Demand Constraint eq)					Warehouse (Supply Constraint ub)		Cost
i to j	1	2	3	4	5	A	B	(USD)
X_{11}	1					1		2
X_{12}		1				1		4
X_{13}			1			1		5
X_{14}				1		1		2
X_{15}					1	1		1
X_{21}	1						1	3
X_{22}		1					1	1
X_{23}			1				1	3
X_{24}				1			1	2
X_{25}					1		1	3
$=, \leq$	500	900	1800	200	700	1000	3200	Z

Source: Table by author.

4.1.2. Solution Exercise

Figure 4.2 shows a graphical representation of the solution. The minimum shipment cost was USD 8600.

Create a copy of the PFT in an Excel file and then run the Python program in the next subsection.

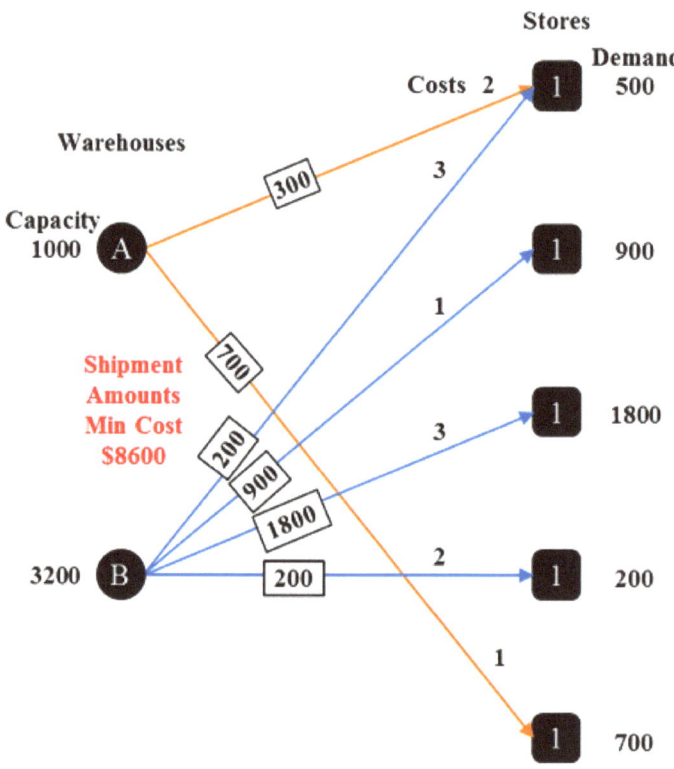

Figure 4.2. Representation of the spatial distribution solution. Source: Figure by author.

4.1.3. Program Display

As with all the examples, be sure to change the variable "datapath_in" to match the path where the Excel file is located. Change the "infile" variable to match the name of the Excel file. The program extracts the PFT by skipping the first row to create a data frame. Because the software imports blank cells as "nan", it replaces these values with zeros. It stores the variable names in the first column as a series of strings. Change the problem scenario to designing capacities for each warehouse to minimize the logistics cost of the system by switching the comments on the inequality constraints. Observe that the total logistics cost of the system drops to USD 8400 after changing the capacities of warehouses A and B to 1400 and 2700, respectively.

```
# Author: Dr. Raj Bridgelall (raj.bridgelall@ndsu.edu)
# Spatial distribution to balance supply and demand
from IPython import get_ipython
get_ipython().magic('clear')                              # Clear the console
get_ipython().run_line_magic('matplotlib', 'inline')      # plot in the iPython console
import pandas as pd
from pathlib import Path
from mip import *         # install library from Anaconda prompt: pip install mip
#%%
datapath_in = 'C:/Users/Admin/Documents/Spatial Logistics and Flows/Lab/'
infile = 'Warehouse Distribution PFT.xlsx'                # Input filename
filepath_in = Path(datapath_in + infile)                  # Path name for untruncated signal
df = pd.DataFrame(pd.read_excel(filepath_in, skiprows=1)) # Read Problem Formulation Table (PFT)
#%%
Nc = df.shape[1]-2      # Number of constraints
Nd = df.shape[0]-1      # Number of decision variables
#%%
df = df.fillna(0)                                 # replace all NaN (blanks in Excel) with zeros
VarName = df.iloc[0:Nd,0]                         # String of variable names
A_df = df.iloc[0:Nd,1:Nc+1].astype(int)           # Extract the constraints cols
A_parms_eq = A_df.iloc[:,0:Nc-2]                  # Extract A_eq
A_parms_ub = A_df.iloc[:,Nc-2:]                   # Extract A_ub (already in standard ub form)
b_parms = df.values[-1][1:-1].astype(int)         # Get constraint parameters in an array
b_parms_eq = b_parms[0:Nc-2]                      # Extract b_eq
b_parms_ub = b_parms[Nc-2:]                       # Extract b_ub
c_parms = df.values.transpose()[-1][:-1].astype(int)  # Get objective parameters in an array
#%%
m = Model(solver_name=CBC)                        # use GRB for Gurobi
# define list of decision vars plus store in x
x = [m.add_var(name=VarName[i],var_type=INTEGER) for i in range(Nd)]
m.objective = minimize(xsum(c_parms[i]*x[i] for i in range(Nd))) # add objective function
#%% Add each constraint column
for j in range(len(b_parms_eq)):
    m.add_constr( xsum(A_parms_eq.iloc[i,j]*x[i] for i in range(Nd)) == b_parms_eq[j],\
                 "ConsD_"+str(j) )
for j in range(len(b_parms_ub)):
#    m.add_constr( xsum(A_parms_ub.iloc[i,j]*x[i] for i in range(Nd)) <= sum(b_parms_eq),\
#                 "Cons_AnyCap_"+str(j) )
    m.add_constr( xsum(A_parms_ub.iloc[i,j]*x[i] for i in range(Nd)) <= b_parms_ub[j],\
                 "Cons_Cap_"+str(j) )
#%%
Status = m.optimize()
#%% Print the results
print('Model has {} vars, {} constraints and {} nzs'.format(m.num_cols, m.num_rows, m.num_nz))
print("Number of Solutions = ", m.num_solutions)
print("Status = ", Status)
Arcs = [VarName[i] for i in range(Nd) if x[i].x != 0]
ArcVals = [x[i].x for i in range(Nd) if x[i].x != 0]
for i in range(len(Arcs)):
    print('{} = {} '.format(Arcs[i], ArcVals[i]))

print('Minimum Cost = ${}'.format(m.objective_value))
print('Constraints:')
for j in range(Nc):
    print(m.constrs[j])
```

4.2. Flow Maximization

This type of problem optimizes flows through a network for a set of origins and destinations. Typical applications in transportation include the following:

- **Evacuation routing**: The constraints of each arc in a path can be the roadway capacity in maximum vehicle volume per hour based on the speed limit, traffic control devices, and geometric features of the roadways (number of lanes, shoulder width, medians, terrain, interchange density, etc.). The traffic source nodes can be towns affected by a threat, and the sink nodes can be candidate towns that can serve as temporary sanctuaries.
- **Pipeline network throughput**: The constraints of each arc in a path can be the flow capacity in maximum flow units per day (such as barrels of oil). The source nodes can be oil production regions such as the Bakken shale, and the sink nodes can be refineries or reservoirs.
- **Railroad network throughput**: The constraints of each arc in a path can be the track capacity in maximum flow units per day (such as shipping containers). The source nodes can be shipping ports or terminals, and the sink nodes can be transshipment terminals that load trucks for delivery to warehouses or stores.

The links considered, indexed as k, are candidate routes between nodes. The node labels and constraints that define the network connectivity and flow conservation at each node are similar to those from the shortest path problem. However, there are four major differences:

(1) The objective function defines the source node s by maximizing flow out of it.
(2) The constraints for flow conservation define the network and the node that has no exit flow as the sink node t without a capacity constraint.
(3) The decision variables are real numbers, not binary values.
(4) Each decision variable has an upper bound that is equal to its flow capacity.

The optimization problem is expressed as follows:

Maximize

$$F = \sum_{j=1}^{N} X_{sj} \tag{53}$$

which is subject to

$$\sum_{j=1}^{N} X_{ji} - \sum_{j=1}^{N} X_{ij} = \begin{cases} -F & \text{if } i = s \\ F & \text{if } i = t \\ 0 & \text{otherwise} \end{cases} \tag{54}$$

where

$$0 \leq X_{ij} \leq U_k \tag{55}$$

The variable N is the number of input or output links at each node. The last column of the PFT contains the upper bounds for each of the decision variables. The lower bound of zero is the same for all the variables, so they need not be in the PFT.

4.2.1. Example Problem

This example uses the same cities as the shortest path problem, but the arcs represent railroad service. The source node is a seaport in Los Angeles. The sink node is a transshipment terminal in St. Louis. Table 4.3 is the PFT that corresponds to this example. The capacities are in units of 100,000 containers per day.

Table 4.3. PFT for the maximum flow problem exercise.

Available Links	Network Flow Constraints					Maximize Flow	Capacity U_k
	1	2	3	4	5		
X_{12}	1					1	5
X_{13}		1				1	3
X_{14}			1			1	2
X_{24}	−1		1				5
X_{25}	−1			1			3
X_{34}		−1	1				5
X_{36}		−1			1		3
X_{45}			−1	1			1
X_{46}			−1		1		3
X_{47}			−1				4
X_{57}				−1			5
X_{67}					−1		1
=	0	0	0	0	0	F	

Source: Table by author.

The objective function is the *dot product* of the decision variables and the column next to the last column such that the optimization problem takes the following form:

Maximize
$$F = X_{12} + X_{13} + X_{14} \tag{56}$$

which is subject to
$$X_{12} - X_{24} - X_{25} = 0 \tag{57}$$

$$X_{13} - X_{34} - X_{36} = 0 \tag{58}$$

$$X_{14} + X_{24} + X_{34} - X_{45} - X_{46} - X_{47} = 0 \tag{59}$$

$$X_{36} + X_{46} - X_{57} = 0 \tag{60}$$

$$X_{36} + X_{46} - X_{67} = 0 \tag{61}$$

where
$$0 \leq X_{ij} \leq U_k \tag{62}$$

4.2.2. Solution Exercise

The solution, illustrated in Figure 4.3, is that trains can haul 900,000 containers per day from the port in Los Angeles. The arc values in black and red font represent the flow volume and the link capacity, respectively. There are currently no constraints on the number of containers that the terminal facilities in St. Louis can receive. However, one can be included in this problem by adding the following constraint:

$$X_{57} + X_{47} + X_{67} \leq U \tag{63}$$

with U being the upper-bound capacity.

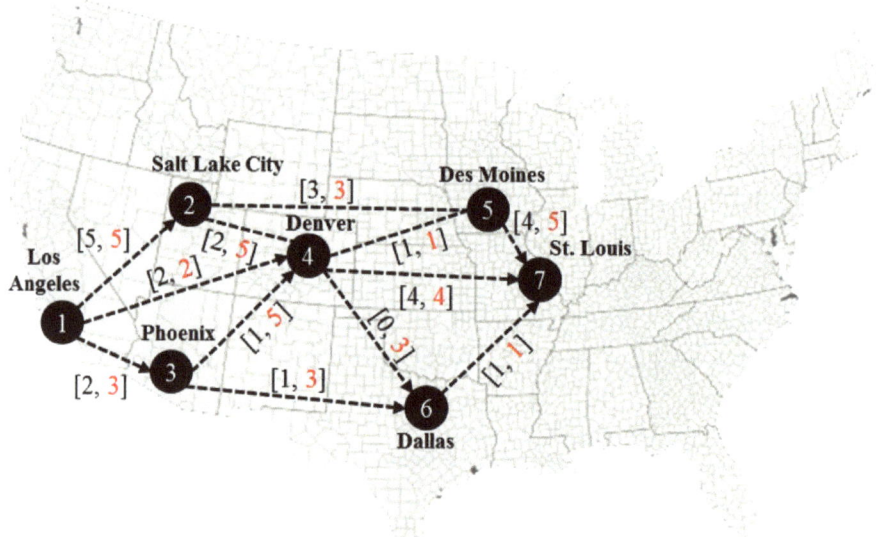

Figure 4.3. Solution to the example maximum flow problem. Source: Figure by author.

It turns out that this solution is not unique, but the maximum flow is unchanged. Figure 4.4 illustrates an alternate solution.

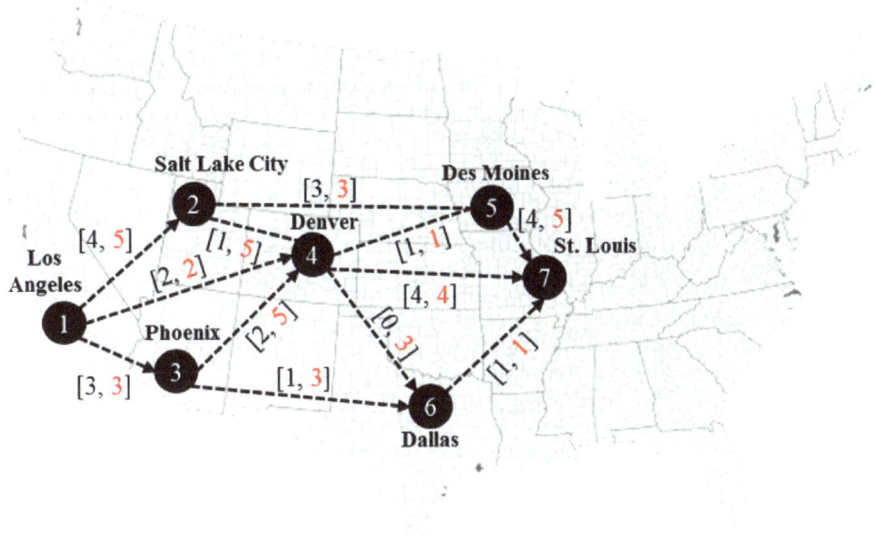

Figure 4.4. An alternate solution to the example maximum flow problem. Source: Figure by author.

4.2.3. Program Display

```
# Author: Dr. Raj Bridgelall (raj.bridgelall@ndsu.edu)
# Maximum Flow Optimization using scipy.optimize
from IPython import get_ipython
get_ipython().magic('clear')                           # Clear the console
get_ipython().run_line_magic('matplotlib', 'inline')   # plot in the iPython console
import pandas as pd
from pathlib import Path
from scipy.optimize import linprog
import numpy as np
#%%
datapath_in = 'C:/Users/Admin/Documents/Spatial Logistics and Flows/Lab/'
infile = 'Maximum Flow PFT.xlsx'                       # Input filename
filepath_in = Path(datapath_in + infile)               # Path name for untruncated signal
df = pd.DataFrame(pd.read_excel(filepath_in, skiprows=1))  # Problem Formulation
#%%
Nc = df.shape[1]-3     # Number of flow constraints
Nd = df.shape[0]-1     # Number of decision variables
#%%
df = df.fillna(0)                                      # replace all NaN (blanks in Excel) with zeros
A_df = df.iloc[0:Nd,1:Nc+1].astype(int)                # Extract the A parameters
A_parms_eq = A_df.values.transpose()                   # Convert the A parametersto a matrix but transposed
b_parms_eq = df.values[-1][1:Nc+1].astype(int)         # Get constraint parameters in an array
#%%
c_parms = df.values.transpose()[-2][:-1].astype(int)   # Get objective parameters in an array
c_parms = c_parms * -1                                 # Transform to a maximization problem
#%%
x_ub = df.values.transpose()[-1][:-1].astype(float)    # List of x upper bounds
x_lb = [0] * len(x_ub)                                 # List of lower bounds (zeros)
#%%
x_bounds = list(zip(x_lb, x_ub))                       # list of lower and upper bound tuples
#%% Call the optimizer
# https://docs.scipy.org/doc/scipy/reference/generated/scipy.optimize.linprog.html
res = linprog(c_parms, A_eq=A_parms_eq, b_eq=b_parms_eq,
              bounds=x_bounds,
              method='revised simplex')
#%% Print the results
print(res); print("\n")
Sol = res.x.astype(int)
print("Values for Decision Variables:\n", Sol)
Arcs = [df.values.transpose()[0][i] for i in range(Nd) if Sol[i] > 0]
Arc_Vals = [Sol[i] for i in range(Nd) if Sol[i] > 0]
print('\n(Arc, Flow):\n', list(zip(Arcs, Arc_Vals)))
print("\nMaximum Flow = ", abs(np.dot(Sol,c_parms)))
```

4.3. Warehouse Location Optimization

Some problems require a mix of continuous and binary decision variables. One such problem is the *facility-locating problem*, where cost optimization is based on both a per-unit cost and a fixed cost. The per-unit cost includes the costs of producing an item plus transporting the item from facility i to consumer j. A fixed cost arises if a facility is open and incurs operating, maintenance, rental, and insurance costs.

The general problem is as follows: given N facilities and M customers, determine which facilities to open and how these facilities can satisfy the demand from all the customers while minimizing cost. The optimization problem is expressed as follows:

Minimize

$$Z = \sum_{i=1}^{N}\sum_{j=1}^{M} c_{ij} Y_{ij} + \sum_{i=1}^{N} f_i X_i \tag{64}$$

which is subject to

$$\sum_{i=1}^{N} Y_{ij} = d_j (j = 1, 2, \ldots, M) \tag{65}$$

and

$$\sum_{j=1}^{M} Y_{ij} \leq u_i X_i (i = 1, 2, \ldots, N) \tag{66}$$

where

$$Y_{ij} \geq 0 (i = 1, 2, \ldots, N), (j = 1, 2, \ldots, M) \tag{67}$$

$$X_i \in \{0, 1\} \quad (i = 1, 2, \ldots, N) \tag{68}$$

The first part of the objective function is the total variable cost based on the per-unit costs c_{ij} for producing and transporting items from warehouse i to store j. The second part of the objective function is the total of the fixed costs f_i for keeping warehouse i open. The decision variable Y_{ij} is the number of units that warehouse i can supply to store j, where the demand is d_j units. Hence, the variables are integers with a lower bound of 0. The second constraint added by Equation (66) ensures that the total units shipped from a warehouse is no more than the supply capacity u_i of said warehouse. Hence, if it is decided that a warehouse will be closed, this constraint ensures that all Y_{ij} values for this warehouse will be zero.

Table 4.4 shows the PFT for the warehouse location problem. The last row of the demand constraints contains the number of units of demand d_j for store j for a given period, such as one day or one week. The last four rows of the cost column contain the fixed cost f_i of keeping warehouse i open for the same period. The remaining cells of the cost column store the per-unit cost for producing and transporting items from warehouse i to store j.

Table 4.4. PFT for the warehouse location problem.

Ship i to j	Stores (Demand Constraint eq)					Warehouse (Supply Constraint ub)				Cost (USD)
	1	2	3	4	5	1	2	3	4	
Y_{11}	1					1				1
Y_{12}		1				1				2
Y_{13}			1			1				3
Y_{14}				1		1				4
Y_{15}					1	1				5
Y_{21}	1						1			5
Y_{22}		1					1			4
Y_{23}			1				1			3
Y_{24}				1			1			2
Y_{25}					1		1			1
Y_{31}	1							1		1
Y_{32}		1						1		2
Y_{33}			1					1		3
Y_{34}				1				1		4
Y_{35}					1			1		5
Y_{41}	1								1	5
Y_{42}		1							1	4
Y_{43}			1						1	3
Y_{44}				1					1	2
Y_{45}					1				1	1
X_1						−60				20
X_2							−10			30
X_3								−50		20
X_4									−55	30
=, ≤	10	20	30	40	50	0	0	0	0	Z

Source: Table by author.

4.3.1. Solution Exercise

The solution to satisfying the demand for the example problem is expressed as follows:

$Y_{11} = 5.0$
$Y_{12} = 20.0$
$Y_{14} = 35.0$
$Y_{31} = 5.0$
$Y_{33} = 30.0$
$Y_{44} = 5.0$
$Y_{45} = 50.0$
$X_1 = 1.0$
$X_2 = 0.0$
$X_3 = 1.0$

$X_4 = 1.0$

Hence, the solution is to close the second warehouse and incur a minimum cost of USD 410 for all suppliers within the referenced period.

4.3.2. Program Display

```python
from pathlib import Path
from mip import *         # install library from Anaconda prompt: pip install mip
import numpy as np
#%%
datapath_in = 'C:/Users/Admin/DocumentsTL 885/Lectures/10 - Spatial Logistics and Flows/Lab/'
infile = 'Warehouse Location PFT.xlsx'            # Input filename
filepath_in = Path(datapath_in + infile)          # Path name for untruncated signal
df = pd.DataFrame(pd.read_excel(filepath_in, skiprows=1)) # Read Problem Formulation Table (PFT)
#%%
Ncd = df.iloc[-1,0]              # Number of demand constraints
Ncs = df.shape[1]-2-Ncd          # Number of supply constraints
Nd_C = Ncd * Ncs                 # Number of continuous decision variables
Nd_B = Ncs                       # Number of binary decision variables
#%%
df = df.fillna(0)                        # replace all NaN (blanks in Excel) with zeros
Var_C = df.iloc[0:Nd_C,0]                # String of continuous variable names
Var_B = df.iloc[Nd_C:-1,0]               # String of binary variable names
#%%
A_df = df.iloc[0:Nd_C+Nd_B,1:Ncd+Ncs+1]          # Extract the constraints cols
A_parms_eq = A_df.iloc[:,0:Ncd].astype(int)      # Extract A_eq
A_parms_ub = A_df.iloc[:,Ncd:].astype(float)     # Extract A_ub (already in standard ub form)
b_parms = list(df.values[-1][1:Ncd+1].astype(int))  # Get constraint parameters in an array
c_parms = df.values.transpose()[-1][:-1].astype(int) # Get objective parameters in an array
#%%
m = Model(solver_name=CBC)                       # use GRB for Gurobi
y = [m.add_var(name=Var_C[i],var_type=INTEGER) for i in range(Nd_C)]  # decision vars y
x = [m.add_var(name=Var_B[i],var_type=BINARY) for i in range(Nd_C,Nd_C+Nd_B)]  # decision vars x
yx = y + x                                       # combined list of decision variables
#%%
m.objective = minimize(xsum(c_parms[i]*yx[i] for i in range(Nd_C+Nd_B))) # add objective function
#%% Add each constraint column
for j in range(Ncd):
    m.add_constr( xsum(A_parms_eq.iloc[i,j]*yx[i] for i in range(len(yx))) == b_parms[j], "ConsY"+str(j)
)
for j in range(Ncs):
    m.add_constr( xsum(A_parms_ub.iloc[i,j]*yx[i] for i in range(len(yx))) <= 0, "ConsX"+str(j) )
#%%
Status = m.optimize()
#%% Print the results
print('Model has {} vars, {} constraints and {} nzs'.format(m.num_cols, m.num_rows, m.num_nz))
print('Objective Function: \n',m.objective)
print("Number of Solutions = ", m.num_solutions)
print("Status = ", Status)
#%%
ArcsY = [str(m.vars[i]) for i in range(Nd_C) if y[i].x != 0]
ArcYVals = [y[i].x for i in range(Nd_C) if y[i].x != 0]
for i in range(len(ArcsY)):
    print('{} = {} '.format(ArcsY[i], round(ArcYVals[i],3)))
#%%
VarsX = [str(m.vars[i]) for i in range(Nd_C,Nd_C+Nd_B) ]
Xvals = [x[i].x for i in range(Nd_B) ]
for i in range(Nd_B):
    print('{} = {} '.format(VarsX[i], Xvals[i]))
#%%
print('Minimum Cost = ${}'.format(m.objective_value))
print('Constraints:')
for j in range(m.num_rows):
    print(m.constrs[j])
```

5. Linear Programming Relaxation

The solutions demonstrated in the preceding chapters utilized primarily ILP and BIP methods to find exact solutions. However, exceptionally large BIP problems might be computationally infeasible (Cormen et al. 2022). Linear programming problems that utilize continuous variables are often easier and faster to solve using special purpose algorithms like simplex or interior point compared to their integer programming counterparts. Linear programming relaxation (LPR) is a strategy that simplifies integer programming problems by relaxing the integer constraints, allowing the decision variables to assume continuous values. This strategy can simplify solving BIP problems, ensuring that, under certain conditions, the computationally less intensive linear programming solutions directly translate to ILP solutions without further modifications. LPR often yields a lower bound for minimization problems (or an upper bound for maximization problems) that is useful for gauging the performance of heuristic or exact algorithms.

Problems that are amenable to LPR have the property of unimodularity, where the determinant of every square submatrix of the constraint matrix is 0, +1, or −1. This property guarantees that solutions to the LPR will inherently be integers. LPR also allows for the introduction of heuristics to find satisfactory solutions efficiently when traditional methods are too slow or fail to find any solutions. One example is the traveling salesman problem, in which scaling the solution to a large number of cities can lead to an exponential growth in the possible routes. By allowing the consideration of continuous values between 0 and 1 instead of binary decision variables, the model can consider fractional paths. This provides insights into which paths are likely to be important in the optimal solution and can serve as a bound for more complex algorithms designed to find an exact solution. Hence, heuristics offer practical means to approach problems that, due to their scale or complexity, defy exact analytical solutions. By exploiting LPR, heuristics can navigate towards near-optimal solutions with significantly reduced computational demand. The subsections that follow demonstrate how to use a linear programming library to solve the spatial distribution problem by recasting the decision variables as continuous.

5.1. Linear Programming Model

The scipy.optimize.linprog library provides an optimizer that solves linear programming problems in the following form:

$$\min_{x} c^T x \tag{69}$$

such that

$$A_{ub} x \leq b_{ub} \tag{70}$$

$$A_{eq} x = b_{eq} \tag{71}$$

$$l \leq x \leq u \tag{72}$$

where x is a vector of decision variables, c is a vector of objective function coefficients, b_{ub} is a vector of the *upper bound* constants on the right side of the *inequality* constraint equations, b_{eq} is a vector of the equality constants on the right side of the *equality* constraint equations, and l and u are vectors of the lower and upper bounds, respectively, of the associated decision variable in vector x. A program can derive the A_{ub} and A_{eq} matrices by transposing the corresponding A matrices in the PFT. That is, each row of A_{ub} and A_{eq} must contain the coefficients of the linear inequality and linear equality constraints, respectively, on the decision variables in vector x.

Note that the standard form for the inequality constraint (70) requires the constant to be an upper bound (that is, less than or equal to the corresponding value). Hence, the analyst must convert any constraint formulated as a lower bound (that is, greater than or equal to the corresponding value) to an upper bound form by multiplying each side of the inequality by negative one. One limitation of this optimizer is that it has no provisions for directly accommodating binary variables. However, it is sometimes, but not always, possible to simulate binary decision variables by setting the lower bound to 0 and the upper bound to 1.

5.2. Linear Programming Library

The following program solves the spatial distribution problem using the linear programming optimizer. It extracts the constraint parameters into a matrix and transposes this matrix to the required form, namely, A_{eq}. It also extracts the c and b_{eq} parameters into arrays that the optimizer requires. The program stores the bounds for each variable as a tuple with zero as the lower bound and no upper bound. The program then calls the optimizer and passes all the required parameters in their expected format. The "revised simplex" optimization method produces the best results for integer programming. The program produces the results as an object that contains the following elements:

con:	an array of floating-point values of the *equality* constraint gap.
fun:	a floating-point value of the optimal value of the objective function.
message:	a string describing the exit status of the algorithm.
nit:	an integer of the total number of iterations performed in all phases.
slack:	an array of floating-point values of the *inequality* constraint gap.
status:	an integer representing a status where:
	0: Optimization is proceeding nominally.
	1: The iteration limit has been reached.
	2: The problem is infeasible.
	3: The problem is unbounded.
	4: Numerical difficulties have been encountered.
success:	a Boolean that is true when the algorithm terminates successfully.
x:	an array of floating-point values containing the solution for the decision variables.

5.3. Program Display

```python
# Author: Dr. Raj Bridgelall (raj.bridgelall@ndsu.edu)
# Spatial distribution using Scipy optimizer
from IPython import get_ipython
get_ipython().magic('clear')                            # Clear the console
get_ipython().run_line_magic('matplotlib', 'inline')    # plot in the iPython console
import pandas as pd
from pathlib import Path
from scipy.optimize import linprog
import numpy as np
#%%
datapath_in = 'C:/Users/Admin/Documents/Spatial Distribution/Lab/'
infile = 'Warehouse Distribution PFT.xlsx'              # Input filename
filepath_in = Path(datapath_in + infile)                # Path name for untruncated signal
df = pd.DataFrame(pd.read_excel(filepath_in, skiprows=1))  # Problem Formulation
#%%
Nc = df.shape[1]-2      # Number of resource constraints
Nd = df.shape[0]-1      # Number of decision variables
#%%
#%%
df = df.fillna(0)                                   # replace all NaN (blanks in Excel) with zeros
A_df = df.iloc[0:Nd,1:Nc+1].astype(int)             # Extract the A parameters subset
A_parms_eq = A_df.iloc[:,0:Nc-2]                    # Extract A_eq
A_parms_eq = A_parms_eq.values.transpose()          # Convert A parm subset to matrix, transposed
b_parms = df.values[-1][1:-1].astype(int)           # Get constraint parameters in an array
b_parms_eq = b_parms[0:Nc-2]                        # Extract b_eq

A_parms_ub = A_df.iloc[:,Nc-2:]                     # Extract A_ub (already in standard ub form)
A_parms_ub = A_parms_ub.values.transpose()          # Transpose A params to standard form of optimizer
b_parms_ub = b_parms[Nc-2:]                         # Extract b_ub

c_parms = df.values.transpose()[-1][:-1].astype(float)   # Get objective parameters in an array
x_bounds = (0, None)                                # set lower and upper bounds for the variables
#%% Call the optimizer
# https://docs.scipy.org/doc/scipy/reference/generated/scipy.optimize.linprog.html
res = linprog(c_parms, A_ub=A_parms_ub, b_ub=b_parms_ub, A_eq=A_parms_eq, b_eq=b_parms_eq,
              bounds=x_bounds,
              method='revised simplex')
#%% Print the results
print(res); print("\n")
Sol = res.x.astype(int)
print("Values for Decision Variables: ", Sol)
Arcs = [df.values.transpose()[0][i] for i in range(Nd) if Sol[i] > 0]
Arc_Vals = [Sol[i] for i in range(Nd) if Sol[i] > 0]
print("Arcs: ", Arcs)
print("Arc Values: ", Arc_Vals)
print("Minimum Cost ($) = ", np.dot(Sol,c_parms))
print("Validation from LinProg = ", res.fun)
```

6. Summary and Conclusions

This educational guide introduces laboratory exercises in linear programming that feature several important optimization problems pertaining to supply chain management and transport logistics. Chapter 1 introduced the several types of integer programming and presented a cognitive framework with which to formulate the optimization problem by organizing the facts of the problem in a standard manner. The problem formulation table (PFT) helped to identify all the decision variables, their linear relationships with the constraints and objective function, and their value bounds. The PFT becomes impractical for importing an exceptionally large and mostly sparse table for problems that have many constraints and decision variables. Therefore, some of the exercises show how to use the multiple-integer programming (MIP) library in Python to scale the PFT for larger problems. Chapter 1 described how to integrate two free GIS tools into the optimization workflow to visualize solutions to the optimization problems. The remaining chapters cover optimization problems involving mobility, spatial placements, and distributions in a supply chain. The mobility optimization problems determined the shortest path in a network and the minimum cost tour to visit all nodes only once. The spatial optimization problems explained neighborhood coverage with the least number of facilities, capturing the maximum flows in a network, creating non-competing zones, and optimizing service locations. The spatial logistics problems demonstrated distribution optimizations for meeting demands within capacity constraints at the minimum cost, maximizing flows through a capacity-constrained network, and locating warehouses to minimize variable costs and fixed operating costs. Students and practitioners can modify the demonstration code and GIS workflow to solve their own optimization problems.

Funding: This research received no external funding.

Conflicts of Interest: The authors declare no conflict of interest.

References

Anaconda Inc. 2022. Data Science Technology for a Better World. Available online: https://www.anaconda.com/ (accessed on 14 October 2022).
Anselin, Luc. 2017. *The GeoDA Book: Exploring Spatial Data*. Chicago: GeoDa Press LLC.
Anselin, Luc, Ibnu Syabri, and Youngihn Kho. 2010. GeoDa: An introduction to spatial data analysis. In *Handbook of Applied Spatial Analysis*. Edited by Fischer Manfred and Getis Arthur. Berlin/Heidelberg: Springer, pp. 73–89. [CrossRef]
Brewer, Cynthia, and Mark Harrower. 2020. *ColorBrewer 2.0 Color Advice for Cartography*. State College: Pennsylvania State University. Available online: http://colorbrewer2.org/ (accessed on 24 January 2020).
Cormen, Thomas H., Charles E. Leiserson, Ronald L. Rivest, and Clifford Stein. 2022. *Introduction to Algorithms*, 4th ed. Cambridge: The MIT Press.
GeoDA. 2022. GeoDA: An Introduction to Spatial Data Science. Available online: http://geodacenter.github.io/ (accessed on 14 October 2022).
Hillier, Frederick S., and Gerald J. Lieberman. 2024. *Introduction to Operations Research*, 12th ed. New York: McGraw Hill.
Miller, Harvey J., and Shih-Lung Shaw. 2015. Geographic information systems for transportation in the 21st century. *Geography Compass* 9: 180–89. [CrossRef]

Mitchell, Stuart, Anita Kean, Andrew Mason, Michael O'Sullivan, and Antony Phillips. 2020. Optimization with PuLP. Computational Infrastructure for Operations Research (COIN|OR). Available online: https://www.coin-or.org/PuLP/CaseStudies/a_transportation_problem.html (accessed on 6 January 2020).

NACTO. 2018. *Shared Micromobility in the U.S.: 2018*. New York: National Association of City Transportation Officials (NACTO). Available online: https://nacto.org/shared-micromobility-2018/ (accessed on 5 February 2024).

Pataki, Gábor. 2003. Teaching Integer Programming Formulations Using the Traveling Salesman Problem. *Siam Review* 45: 116–23. [CrossRef]

QGIS. 2022. QGIS: A Free and Open Source Geographic Information System. Available online: https://www.qgis.org/en/site/ (accessed on 14 October 2022).

Santos, Haroldo G., and Túlio A. M. Toffolo. 2020. Mixed Integer Linear Programming with Python. *Computational Infrastructure for Operations Research*. Available online: https://buildmedia.readthedocs.org/media/pdf/python-mip/latest/python-mip.pdf (accessed on 5 February 2024).

Shaw, Shih-Lung. 2011. Geographic information systems for transportation—An introduction. *Journal of Transport Geography* 3: 377–78. [CrossRef]

Taccari, Leonardo. 2016. Integer programming formulations for the elementary shortest path problem. *European Journal of Operational Research* 252: 122–30. [CrossRef]

Taylor, Bernard W., III. 2019. *Introduction to Management Science*, 13th ed. New York: Pearson Education Inc.

Teodorovic, Dusan, and Milica Selmic. 2013. Locating Flow-Capturing Facilities in Transportation Networks: A Fuzzy Sets Theory Approach. *International Journal for Traffic and Transport Engineering* 3: 103–11. [CrossRef]

MDPI
St. Alban-Anlage 66
4052 Basel
Switzerland
www.mdpi.com

MDPI Books Editorial Office
E-mail: books@mdpi.com
www.mdpi.com/books

Disclaimer/Publisher's Note: The statements, opinions and data contained in all publications are solely those of the individual author(s) and contributor(s) and not of MDPI and/or the editor(s). MDPI and/or the editor(s) disclaim responsibility for any injury to people or property resulting from any ideas, methods, instructions or products referred to in the content.

www.ingramcontent.com/pod-product-compliance
Lightning Source LLC
LaVergne TN
LVHW072311090526
838202LV00018B/2264